Family and
The Great Outdoors

Family and
The Great Outdoors

Library of Congress Cataloging-in-Publication Data

Names: Turner, Wayne C., 1942- , author.
Title: Family and the great outdoors / by Wayne C. Turner.
Description: Lilburn, GA : Fairmont Press, 2015.
Identifiers: LCCN 2015036465 | ISBN 0881737674 (alk. paper) | ISBN 0881737682
  (electronic) | ISBN 978-1-4987-6304-2 (Taylor & Francis distribution : alk. paper)
Subjects: LCSH: Hunting--Anecdotes. | Fishing--Anecdotes. | Family
  recreation--Anecdotes. | Turner, Wayne C., 1942-
Classification: LCC SK33 .T87 2015 | DDC 639--dc23 LC record available at
http://lccn.loc.gov/2015036465

Family and The Great Outdoors / Wayne C. Turner

Published by The Fairmont Press, Inc.
700 Indian Trail
Lilburn, GA 30047
tel: 770-925-9388; fax: 770-381-9865
http://www.fairmontpress.com

Distributed by Taylor & Francis Group LLC
6000 Broken Sound Parkway NW, Suite 300
Boca Raton, FL 33487, USA
E-mail: orders@crcpress.com

Distributed by Taylor & Francis Group LLC
23-25 Blades Court
Deodar Road
London SW15 2NU, UK
E-mail: uk.tandf@thomsonpublishingservices.co.uk

Printed in the United States of America
10 9 8 7 6 5 4 3 2 1

ISBN 0-88173-767-4 (The Fairmont Press, Inc.)
ISBN 978-1-4987-6304-2 (Taylor & Francis Group LLC)

Contents

Preface

"If you don't like to be alone hunting, you must not like the guy you're with."

Percy L. Turner (Dad) to Wayne Turner (me)
after a day afield, 1955.

"What I am hunting most, while hunting, is myself."

Ortega Y Gasset

"On about the third day of a good hunt, I find Wayne Turner wandering with me."
Me to our boys talking about the value of finding and knowing oneself.

The above quotes talk about the value of hunting* and self-development. When I am stressed, I seek the woods or stream, for it is then that I am most comfortable and at home. Dad emphasized that in my up-bringing, and it has stuck with me for well over 70 years. This book about a life of hunting is dedicated to my mom and dad, brother Ken, my wife Kathy (today is our 46th anniversary), and our sons Travis and Drew for the effort they all put in to help me find and keep me. It is also dedicated to the finest brace of bird dogs a man ever worked with (not sure who owned whom) especially Okie and Annie. Thanks for sharing your lives with me.

"You did not kill the fish only to keep alive and to sell for food, he thought. You killed for pride and because you are a fisherman. You loved him when he was alive and you loved him after. If you love him, it is not a sin to kill him. Or is it more?"

Earnest Hemmingway—The Old Man and the Sea

"To the sportsman the death of the game is not what interests him; that is not his purpose. What interests him is everything that he had to do to achieve that death—that is the hunt. ...To sum it up, one does not hunt in order to kill; on the contrary, one kills in order to have hunted."

Ortega Y. Gasset

---

*I use the term "hunting" to refer to any harvest activity involving nature. It includes but is not limited to hunting, fishing, frog gigging, mushrooming, trot lining, etc. Rather than give a long list each time, I shall say "hunting."

"Hunting is nothing more than pursuing the game and laying hands on it…"

Plato (Socrates and Plato both hunted.)

The above quotes lend credence to my belief that it is the pursuit, the understanding of nature and that animal, fish, or vegetable pursued, not the kill itself that is important. However, I must periodically harvest, or I have not hunted. Even Thoreau and Muir believed that the best way to sensitize youth to nature is to bring them up hunting and fishing. I think they would be disappointed that at 73, I am still hunting and fishing. Perhaps I am still looking for Wayne?

Now, you know me as well as I know me, but I am still exploring. This book is a collection of stories about that search. Part I, "Raised in the Great Outdoors" is about me as a child being raised outdoors as Mom and Dad both believed you should hunt with your child not for your child. This part also includes my meeting and falling in love with Kathy* to whom I have now been married 46 years. This period runs from my birth in 1942, until Kathy and I had kids and moved to Oklahoma in 1974.

Part II, "Raising the Boys in the Great Outdoors" is about our boys, Travis and Drew, and how we attempted to carry on the tradition by raising them hunting in the great outdoors. This part covers the period from 1974 until approximately 2000—the time when the boys were growing up. Raising them was the most rewarding thing Kathy and I have ever done, and you will see in the stories the importance of raising them in the outdoors. Those stories involve Kathy, Travis, Drew, and the finest brace of bird dogs a man ever worked with, and we were all the best of friends. One son continues the tradition, while the other loves the outdoors but chooses not to kill. Both are happy and functioning well as adults so we didn't screw it up too badly.

Part III, "The Old Man Grows Older in the Great Outdoors," is about my final and perhaps most productive leg on that journey of discovery. It covers the period from 2000 until tomorrow. The boys moved away, but today our pursuits continue. You will find a dramatic change in philos-

*While pursuing my PhD, I met and fell in love with Kathy, my wife now of 46 years. Guess what, we both admit that we fell in love fishing. On our first date, a bucket of hellgrammites (ugly black critters with big pinchers) turned over in my car, and I had to recapture them (I didn't want to kill them as they are great bait). She laughed and asked if she could go fishing with me; who wouldn't fall in love with a gal like that? We are still fishing together but along the way, we created and raised two great boys.

ophy, as I realize the kill is much less important than the chase. The story, "The Mystic Side of Nature," was a major discovery in that journey that frankly hit me hard as I made a huge leap in understanding. Today, I always say a prayer over and thank the animal for his/her sacrifice. You'll like that story I think.

Part IV is about a recent trip Kathy and I took to Africa to hunt and photograph. It is perhaps the culmination of my search, although I do intend to go back, and as such, the story has its own separate section. I believe it was Ruark who said, "No one ever goes to Africa once," and yes, I still have the urge.

Most importantly, this book is intended to be fun and show how living outdoors is a blast. I highly recommend it, but that's enough for now; I have crab pots to bait at our place on the Chesapeake Bay, and I must get the clam rakes together for an afternoon jaunt on the flats.

Enjoy. Tight Lines and Drag Free Floats,

Wayne Turner
Grimstead, VA

In our annual archery elk camp, family and close friends gather around the campfire and tell stories. One of the close friends, Marty, is an accomplished artist so he takes a burnt stick and draws the object of the story (as he sees it) on a flat rock. Next night, we clean that rock and do it again. Those stories are in this book and some of the rock art has been transferred to paper. Smell the smoke, hear the laughter, and feel the love. Hopefully this book will help you get there.

Stories' Philosophy

I wrote all these stories for the family over a period of about 35 years. Thus, you will see some differences in style that I chose not to change, for that's who I was at the time. Some of the stories were published in magazines such as *Virginia Wildlife, BaySplash, Lightin' Ridge,* and *Outdoor Oklahoma.* I've always intended to publish a "lifelong experiences in the outdoors" book like this one.

I did employ a few guides or outfitters (Parts II and III), but most experiences were our own, with just family and close friends present. Only once did I trade a hunt for an article, and that was in Part III.

Thus, the vast majority of these stories were written only for family and close friends. Since you are now in my close friends circle, here they are.

Acknowledgment

Sidney Read created all of the artwork in this book—some at the author's request, some from his own response to the stories, and some under the gun of the printer's clock. Thanks, Sidney!

Read Art Studio
Sidney Read
Mathews, Virginia

Sidney's work is on display at Frenchy's Art Gallery in Mathews, Virginia. He can be contacted by phone at 804-384-6220 or by email: sjread10@gmail.com.

About the Author

Wayne Turner lives in Dillon, CO, with his wife, Kathy, who fishes better than him, and a bird dog, Annie, who hunts better than he—nevertheless, both still love him, and he them. Wayne and Kathy have raised two wonderful sons, Drew and Travis, and are proud grandparents. Plus, they've raised a bunch of great bird dogs.

Raised in the swamps of eastern Virginia, Wayne's love of the outdoors was imbued in him by his father, whose hunter-gatherer nature made an indelible impression on Wayne from an early age. His father shaped him into the man he is today, and his lasting influence is evident in three generations of progeny.

As an outdoorsman, he is dedicated to conserving the environment for future generations. He has more shotguns and fly rods than any man should have, and is still collecting. On the Chesapeake Bay, Wayne and Kathy are licensed crabbers, oyster farmers, and avid kayak fisher-persons.

As an "indoorsman", his conservation ethic guided his life's work. Wayne is widely considered the father of modern-day energy management. He co-authored the definitive text on energy management, now in its seventh edition, used in universities around the world. He is a Regents Professor Emeritus at Oklahoma State University, Emeritus Editor of two engineering journals and author of more than 200 professional articles. He is also a member of the Association of Energy Engineers Hall of Fame, and one of the first three Fellows in that Association.

This book reflects on his outdoor life, his family, and his lifelong journey toward finding and knowing himself. This continuous quest brings him closer to that goal every day.

PART I

Raised in the Great Outdoors

Dad, Ducks, and Boy Raising

Alarm Clock

The smell of country ham and eggs frying along with coffee brewing (for Dad, hot chocolate for me) was my alarm that it was time to get up and go duck hunting with Dad. I'd put on my woolens (yes, pure itchy wool with the proverbial back door; if synthetics existed we didn't know it) and join him in the kitchen for our morning ritual of getting ready. I sure wish I could do that again. We'd eat, clean the fry pan (a little) and get into the jeep that he already had running. I'm not sure why, for with the lack of sides and only a tattered top, the heat would never really take effect but it was better than nothing, I guess. We'd drive to the hunt.

The Woodie Hole

About five miles north of West Point, VA, near aptly named "five mile hill," was a small pond loaded with woodies in heavy timber on a creek. I had spent many hours raking, picking up sticks, and generally clearing a path 100 yards so that I could sneak to the hidden pot hole with prearranged logs blocking the view of a young kid (me), shaking like a leaf, sneaking to get within a .410 range of the woodies. I was carrying my first shotgun that Dad had just given me, a .410 ga. Savage over and under. Yes, with that shotgun and its limited range, I was to shoot them in the water, the air, or anyway I could. This was 1952 or so.

Getting to the logs, I peeked over, and yes, they were there, but they swam up the creek led by a gloriously colored drake. Dad not hearing shots, came to me in a half crouch. I explained to him and he said, "stay here." He walked a large semi-circle around the pothole back to the creek far up stream. I heard a loud piercing wood duck screech and sure enough, here they came but they were flying. I think, "do this right" so I sat on the ground, led the drake by about 2 feet and shot that full choked .410 at less than 20 yards. There is a God—the drake fell, and I was the proudest 10-year-old in King William County. We didn't take any pictures and I think the only camera we had was a small Brownie, at least I think that's what we called it. Dad and

I plucked that duck, and Mom fixed it for me that night with gravy over rice—a grand moment. We killed many more ducks over the years but that one was my proudest; Mom and Dad knew that, and if I was present, they bragged about it in front of anyone who would listen. That was just fine with me; I let them brag, and grew a couple inches each time they told the story.

Country Club Mallards

Dad joined the local country club which surprised me a great deal as golf courses were a source of night crawlers, but that was about it for the Turners. Oh, and we caught a lot of big bass in the lake, but I never went in the clubhouse or touched a golf club until many years later. Duck season was getting close, so Dad went to a hidden cove chosen so we shot away from the clubhouse. Once again, we cleared a path, and we snuck up on that pond on the way back from deer hunts in King William County. By then I was shooting a single shot 12 ga. Savage, and I was pretty salty with it, or at least I thought so. This was about 1957 when I was 15 years old, going on 25.

We'd sneak together, both on all fours, getting close to the splashing sound and sometimes quacks that would tell us the ducks were there. Dad

would nod and we'd stand. I'd shoot once, he'd shoot three or so times, and often four or so ducks would lay on the water. Mallards loved that spot, so many fell during the season but, you know, our membership only lasted one year, and Dad never told me why but I suspect that the golfers didn't especially like shotgun sounds when they were teeing. We were pretty close to the actual course, but again, safety was important to Dad and we were shooting away from the clubhouse which was about ¼ mile from us.

Mattaponi Tributaries Floats

Many creeks drain into the Mattaponi and Pamunky Rivers, often through marshes, and they're all tidal. We'd take an old wooden rowboat swollen so that it was pretty water proof and put a small kicker motor on the back. Launching close to the chosen creek, we'd motor to the creek mouth just as the incoming tide made the creek pretty navigable. We'd kill the kicker, and paddle with the current, one lying in the bow and the other paddling in the back. I don't remember laying trees across the bow on these trips, but later I did that to give us a few more seconds of sneak.

The creeks all meandered a great deal, so from start to finish, we probably never went more than ½ mile as the crow flies but what seemed like 2 or 3 miles as the creek meanders. Seldom did we have any warning—we'd round a bend, and a flock of dabblers would jump. Dad would shoot his Browning three or so times and usually three or so ducks would fall. When it was my turn, I'd shoot my single shot once, and sometimes a duck would fall.

Getting them often involved only paddling, but sometimes a trek into the marsh was needed. The tide was flooding pretty strong, so a retriever could easily get them, but then, the duck dog looked a lot like Dad and me. Now I have a yellow lab that would love to get them for me. She actually is a part-time daughter, since the kids have grown and left, and sometimes a retriever/flusher.

Miracle of miracles, about the time we were tired, the tide would turn and we'd do the same coming out. We never shot as many on the return (I guess due to the commotion), so the trip went faster. If I do my math correctly, the incoming tide would take 6 hours and we'd wait for it to flood pretty good so we must have spent 4 hours going in and a couple coming out. This was one of my favorite hunts but, hey, thanks to a great dad, I had MANY favorite hunts.

Cans on the Point

Our home, West Point, was a peninsula bordered by the Mattaponi and Pamunky Rivers that formed the York River at the end of the point. Dad always had some judiciously placed decoys around docks on the Point (usually in the York or Mattaponi Rivers). Canvasback ducks ("Cans") loved that area and we would frequently sneak up on the decoys on our way to another hunt. Yes, sometimes homes were near, but we always had permission, and we always shared or offered to share the meat.

Canvasback ducks are tasty critters and highly coveted then and today. Read the stories of market gunning days, and you will see that Cans brought the most money and were the easiest to sell. We never once sold any, but we sure ate a bunch and gave many to friends.

We'd first scout and make sure there were ducks, which was often done from an old dock farther up the Mattaponi. Then, we'd sneak through the marsh using marsh grass and docks as sneaking cover. Cans and sometimes redheads would flush, and we'd shoot. The shots were typically pretty long (40 yards or so) but we'd still kill quite a few. By then, I was shooting an 1148 Remington 12 ga semiautomatic with a full choke. In those days, the folks

would often walk out to see how we did and we'd give them a few.

Dad

By now, you understand, I had the greatest dad in the world; he and I had a relationship that they write about (guess I just did) and I miss him every day of my life. These hunts started about 60 years ago and ended about 40 years ago. In a way, they're ongoing, as I still hunt with Dad through memories.

You should know, I did the same for our two boys and we still hunt/fish together.

Dad had moved to West Point from Appalachia during the depression to work at the local paper mill. He dropped out of school at 12 years old to help support the family, and he hunted frequently to help feed them. He never lost the urge or need to bring home meat, sometimes lots of it. He ran trap lines for spending money and hunted about anything that was tasty.

Thanks, Dad.

Lunch at the Crossroads

Up King William County about halfway from West Point to King William High School, there was a crossroads with a small general store on the southeast corner. We knew the owner well and he was always ready to fix lunch for us. We'd have country ham with mustard and sweet pickles on (of course) white bread. Grabbing an RC Cola, we'd go to the potbellied stove in the corner with a couple chairs pulled near the heat where the owner would join us. That stove was always really hot when we got there, and this was about my favorite time of the hunts. The store was near all the hunts described in this article except for the Mattaponi float. After the sandwich, we'd grab a Moon Pie (honestly) which with the RC made quite a treat to this kid. Many problems were solved or set aside around that stove, including some rather urgent puberty-induced issues involving a teenager with a large libido and a dad that understood.

Reluctantly, we'd get up and go back to the hunt or home. Dad knew how important such treats were to a young kid, and he was correct; I will never forget our trips to the crossroads store.

Wroblick's Huntin' House

Previously published in *Virginia Wildlife*

Start in Richmond, Virginia, or thereabouts and head east in a car. When the air gets thick and humid and smells of salt and marsh (ignore the paper mill smell), you're getting close. When you get to West Point, go north almost 5 miles and turn east. When the tires quit whining on asphalt and start squish/squishing in sand, you're almost there. Stop, get out, and continue walking east. When you're about ankle deep in water, mud, and muck, go another few yards and you're there!

Once you get there, glance around a bit; those rusted, paper-hulled shotgun shells might have been mine. I fired many of them through that old break action, single-shot, 12-ga Savage Arms shotgun and, to brag a bit, every once in while they brought that duck, rabbit, etc. down. There are many stories about frog gigging, duck huntin', and trappin' in those marshes, but this one is about deer huntin' around my home town of West Point, Virginia.

We belonged to a hunt club that we'll call the Wroblick's Club. Meeting every hunt day (which was very often during season), we'd get together while still dark in the Wroblick's Huntin' House. At one time, the house was white—I could tell from the bits of white paint stubbornly hanging onto the weathered siding. The house was small, but it had one rather large room that was the gathering site for the day.

Close your eyes and walk with me into that room. Let your sense of smell and hearing guide the way. First, your nose picks up memory generating scents of wet wool (synthetics didn't exist), wet dog fur, and gun-cleaning solvent. Unfortunately, there was also the scent of tobacci juice spat into empty coffee cans sitting on the wood stove (ok, there were some unpleasing parts). Then, your hearing detects wind rattling around loose window panes in that old house. You also hear the wood stove sucking in air as it heats the house, coffee boiling on the wood stove and men telling stories to each other. Open your eyes and you'll see about 10 old men sitting around, and one sandy-haired kid curled around the wood stove. That kid is yours truly about 70 years ago. This was one of my favorite times of the day. Stories that everyone knew were only partially true were flying everywhere. That's also where I learned to curse, much to Dad's dismay.

We hunted paper mill lands or local farms. All of them contained significant swamps, and the difficulty of hunting in the ever-present swamps was our argument for using dogs to chase the deer out to where we could get to them. That practice and those packs of hounds define a whole culture that was extremely important to me then and still is to many. Today, I choose to hunt my deer with a recurve bow and black powder rifle, but I would fight to save that culture. Somewhere today, there's another huntin' house or two with kids in it that will become the next Archibald Rutledge, Havila Babcock, Sigurd Olsson, Adolph Leopold, etc., of our outdoor/naturalist literature. Even Henry David Thoreau said the best way to raise a naturalist is to start him/her hunting and fishing.

The day you and I are visiting is the day we had chosen to hunt Olson's farm, so we loaded the dogs in the boxes and drove to the site. After parking the trucks, the dog boxes were opened and out climbed about a dozen dogs. There are Walkers, Blueticks, Black and Tans, Redbones, and about every combination/permutation thereof. Ancestral lineage was often determined by happenstance rather than careful planning.

There'd be a three-legged dog in the bunch along with a one-eyed hound (sometimes the same one), and many of the dogs would carry battle scars. Dogs would run around everywhere emptying bladders and bowels. To this day, I'm impressed with the productivity of a hound dog that can eat one pound of food, turn it into three pounds of poop, and still get enough nutrition to run deer for several hours. After the running around was over, I'd get on one knee, and the dogs, knowing this kid was good for an ear rub and maybe even a butt scratch, would come to me. They were among my best friends, and they knew it.

We (people) would scatter around the farms on old farm roads. The farms might run from 100 acres to several thousand. One driver would start the dogs. He would bellow in the swamp, sounding more like a dog than the dogs themselves, but it was easy to tell him from the rest by the skill with which he cursed. Often, the cursing was directed at individual dogs that didn't think running deer was the number-one priority of the day. Usually those dogs didn't stick around long.

Soon, one of the dogs would bark *cold trail*. Over the next 30 minutes to several hours, depending on scenting conditions and IQ of the deer, that cold trail bark would turn into a *hot trail bark*, then a *running bark* (the deer is up and running), and finally a *running by sight bark*. The melody of a dozen hounds running a deer or coon by sight is one of the most thrilling sounds ever heard. That sound alone was worth the trip (read *Where the Red Fern Grows*). Yes, you can tell all those sounds after awhile and even which dogs are making them. That's an important part of the culture.

The sound of a deer running is very distinctive and unmistakable. That day, I heard the deer coming toward me. Down on one knee (to see under the trees better), I was looking hard for the deer but he ran on by me (probably heard my racing heartbeat) down toward my father. Dad visibly stiffened, put his old square forearmed Browning to his shoulder, and I saw smoke. A few seconds later, I heard the shot. Dad lowered the gun and walked into the woods; not many deer ever escaped if he shot.

Paying homage to the fallen animal is something I pride myself on today; but in those days, we'd gut the deer, put him in the truck and go do it again. At the end of the day, we'd go back to Wroblick's and skin him.

There was a skinning tree and for bad weather a skinning shed. See the sketch on the following page, but don't hold me to accuracy, as many years have passed. Of course, this part was very social, and the meat was shared among all parties.

This is when I learned that hamburger or steak was once a living brown-eyed creature, and not something born in a plastic package at the grocery store. Because of this upbringing, I never was bothered with gore and blood. I'd be handy to have around if you ever got hurt, and as soon as the hunt was over, I'd help you, regardless of how much blood.

We'd go home with our meat, put it in a freezer, eat supper, relive the day, and I'd go to bed. Sometimes, it'd still be daylight, but most times, I'd last until after supper and sunset. The next day, we'd do it again.

That's the way it was, and that's the way it should still be.

Thanks Steve, John, Joe, Harry, Ken, Mom, and especially Dad.

In the picture, Dad is standing on the left of the deer, holding an antler (check out those brow tines). Also in the picture are my two grandfathers, standing on the far left (I think). Grandpa Connelly (my mother's side) always carried a lawn chair for deer drives, and he made a great wing bone turkey call. I'd pay a bunch of money for one of his today. Grandpa Turner shot a double barrel Davis that I still own. It now has a new stock (had to do it); but I don't use it, for old time's sake. I do fondle it periodically and always keep it cleaned and oiled.

The Wroblick family at the time consisted of (my best memory) a mother and father who came over from Poland and spoke only broken English and three sons Steve, Joe, and John. Mom and Dad Wroblick made their living from vegetable farming, helping with the deer hunts, and by cleaning and repairing floating seine nets that were used in the spring to harvest spawning shad (whites and hickories) and anadromonus stripers. The sons (along with Dad and other locals) organized the hunts. John often wore riding pants and leather boots (see picture far right standing), Joe ran a local diner when he wasn't hunting (lower right holding dog), and Steve (kneeling between Joe and the deer). Steve and Dad worked at the local paper mill and were inseparable during hunting season.

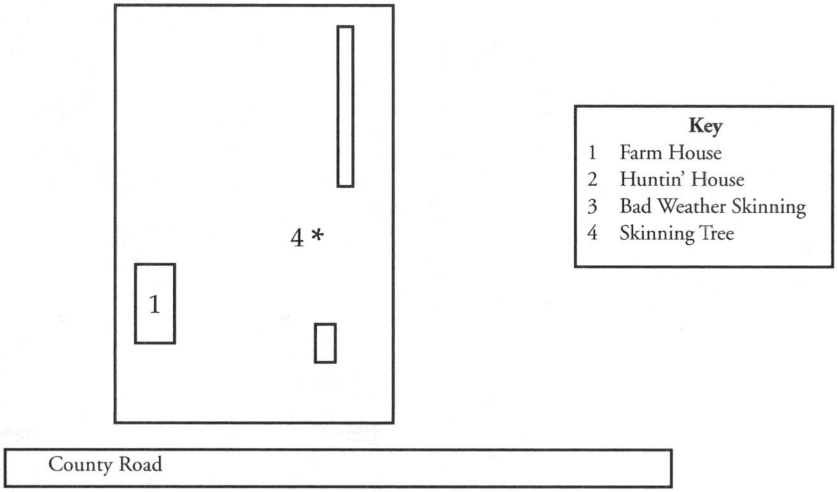

Key

1   Farm House
2   Huntin' House
3   Bad Weather Skinning
4   Skinning Tree

4 *

1

County Road

Wroblick Farm

The Rusty Tackle Box and
A Big Bass

In a corner of the garage was a gold mine for a 9-year-old kid. A rusty tackle box lay ignored, inviting me to open it; so I did. I pried the lid open, and there inside were five or six wooden fishing lures. As I remember, there was an oversized black jitterbug, a huge hula popper, a couple of past-recovery spinners, and a very large crazy crawler. The crazy crawler caught my eye.

Dad and I were primarily Chesapeake Bay fishermen for large croakers and had not done much bass fishing. We had never bass plugged before, since all our previous bass fishing trips involved cane poles, big hooks, and minnows large enough to clean if we came home without a largemouth. I had caught some nice bass with those, but a huge ole bucketmouth had eluded me. I asked Dad what those big plugs were for, and his simple answer was, "big bass." When asked how to use them, he reached into another pile and pulled out an old steel bass plugging rod. It was heavy and rusty. He then produced an old level wind reel with some thick black fishing line on it.

We worked on that reel awhile and finally got it to where with effort we could turn the handle. Lots of oil inside and on the mechanism made it work passably. He then mounted the reel to the rod, tied on a plug, and showed me how to cast it.

I proceeded to demonstrate how easily a young boy can turn that nice ball of line on the reel into a practically impossible-to-fix "birds nest." Next instruction was on how to untangle a bird's nest. Pulling the line out the face of the reel in all sorts of directions until the mess was uncovered and then slowly trying to wind it back neatly on the reel was all that was necessary, but it seemed like rocket science to me.

The rest of that day until late afternoon was spent casting, untangling bird's nests, and learning about bass fishing. Note how willing Dad was to stop everything and devote the afternoon to the important task of teaching his boy how to fish. Remind me to tell you about finding the old Victor traps and what that led to; but that's another story.

After practicing so long, I guess he felt like he had to take me fishing. So, he said something like, "Want to try that out?" Of course, I responded in the affirmative, so we piled into the old red World War II surplus jeep. Top speed was about 40 miles per hour, and that required a downhill head start. We were a team: Dad, me, that old rod and reel with the crazy crawler tied on, the tackle box (in case we wanted to change lures), and my best buddy "Spec" who was a great pointer bird dog that loved ice cream and being with me. Not having any doors and what could only jokingly be called a top, the jeep was bouncy and chilly, but the three of us loved riding in it. Dad drove us over to King and Queen County to an old mill pond. (In eastern Virginia, "mill ponds" were actually small lakes that used to support grinding mills at the stream outlet. In those days, some of the old mills actually still stood but most weren't being used).

This mill pond was about half way up the county, and the road crossed the dam. We used to duck hunt the pond, and I guess we had permission to fish. In those days, getting permission was not difficult, and we did try hard to take good care of the land. The jeep lurched to a stop, we walked about 50 yards to a clear spot, and he said "Go get 'em."

I started casting. By then, I could sometimes make three or so casts in a row without a bird's nest. Thus, I would cast a few times, spend considerable time untangling, and cast a few more times. Note, that I was doing all the casting, repairing, and deciding where to fish. If I asked, he would give advice. Otherwise, he would sit on a log, watch me intently, and (now, I know) pray silently that a bass would hit, while knowing that with this commotion the odds of catching a bass were slim. Simply stated, between Spec wanting to chase the plug, the noise of my yelling at him, and my making bad casts, the odds were not very good.

After maybe 45 minutes of this, I was ready to go home but Dad made me stick with it a bit longer. So, I was casting somewhat dejectedly and retrieving absent mindedly. All of a sudden I heard a loud splash and the reel handle was ripped from my hand. A world-record bass was on. I couldn't grab the handle, so I was screaming for Dad to help, and he was now standing by the log laughing. Finally, I determined that, by gosh I had worked too hard and was going to catch this thing.

I remember jamming my hand into the reel handle so my hand became a brake lever (that hurt). Since the line when new was capable of pulling the jeep (Dad didn't believe in ultra-light fishing, and we'd never heard of it anyway), line breakage was not a concern. Now, I had that sucker; but the metal rod was going crazy telegraphing every move of that bass to my tired arms and shoulders.

I managed to grab the handle and tried to reel. I remember reeling in one or two turns and watching the reel handle burst out of my grip, turning the other way for three or four turns. "Drag" was something we never heard of and maybe that's why they made thick line. At this point, tears showed up on my cheek; I couldn't lose this fish. I turned to Dad and said something like, "I can't do it; please help me." He wouldn't, and he said it was up to me. (That was a statement that I was to hear many times in many different places).

After what felt like two days of hard battle (actually maybe 10 or 15 minutes), the bass started tiring. I could hang onto the handle and actually make progress. I remember seeing this huge bass roll on its side and come sliding up the bank. I think I must have run up the bank dragging it for I remember running back down the bank, grabbing it by the mouth, and pulling it up on shore with that large green crazy crawler dangling out the side. Dad still did not help, but Spec was sure trying by biting the fish.

Not even having heard about "catch and release," and wanting to treasure the moment anyway, Dad helped me clean the fish. I would have been very dangerous with a sharp knife for, by then, I was shaking from head to toe, actually crying at times, and feeling like the biggest MAN that ever existed. Dad was good about making me feel that way. Emerson described this ecstasy as "nature certifying the supernatural, body overflowed by life." Well, mine was!

We started the drive home. I remember laying the fish carefully on the floor on a sheet of paper so it wouldn't get dirty. Spec was very interested and would have loved to have gotten hold of that fish, so I had to watch him carefully. We got home, and I guess we ate it; but, all I remember was telling anyone who would listen what a great deed I had done.

Now, over 60 years later, I still remember that feeling, and I remember that bass as if it were yesterday. There have been many larger bass feel my hook and fly since then. Most have been released, but a few have been killed and eaten. I know that many of them weighed more than that first bass, but none was as big, nor will there ever be a bigger one.

The mere whisper of a reason to go outside and hunt/fish/trap was enough to make Dad drop everything else and go do it. I was fortunate in that if he did it, I did it. He felt part of his job was to bring me up outside enjoying all that the Chesapeake and surrounding area provided. His devotion to spending time with me was phenomenal and rare even in those days—almost unheard of today. Fathers (and mothers), spent time outdoors with the kids; we all remember and appreciate the time together. The Chesapeake and surrounding lands offered (offer?) a virtual smorgasbord of opportunities from fishing, hunting, trapping, clamming, crabbing, to seine netting and more.

Thanks, Dad. I love you and didn't say it nearly often enough.

Deer-dogs and Me
All Pot Lickers

They taught me that activity and heart
can forestall growing old and postpone death,
that the end should come before the stretch,
and that you can stay young and love hard right to the end.
We don't have to grow old and crotchety.

Three score and 5 short years ago, Percy started his towheaded son deer hunting with a pack of pot-licking dogs. The kid began an odyssey that continues today—a search for himself. That kid was me, and those dogs became my best friends. They loved two things—me and chasing deer—so I joined the pack with my mean-shooting Red Ryder. Many deer fell to my rapidly firing BB gun while standing by Dad who fired his Browning 12 ga as back-up. At about 10 years old I moved up to a single shot 12 ga, and officially became serious, but I still loved those dogs.

The pack changed membership constantly, and they were of mixed lineage, but all had sound genetics somewhere back in time, as one of the human leaders would decide the pack needed more mouth, more cold trailing, or faster closing, and would go out and actually pay for a dog. Often, the price was a whopping $50 or so, as they reached out of state and bought a purebred Bluetick, Redbone, Black and Tan, Walker, or some other powerfully bred dog. Perhaps there were fox hounds in there also, but in those days I just didn't know the difference. After the purchase and a planned mating or two, the dog would fall in with the pack, and future genetics were determined by happenstance and fighting ability. Pregnant bitches would still run deer until the humans felt time was near. In spite of all this, I don't ever remember one of them growling at me, so I guess I was accepted into the pack.

What a sight we were! Some of the dogs were sound of frame, but most had scars of some sort. There'd be a three-legged dog or two, a dog with shredded ears that a boar coon or other dog managed to tear, a one-eyed dog that resulted from a poke in the eye whilst running through brambles, and a dog with bent legs (that dog was rather short while standing but didn't lose height while sitting). Before the hunt, we'd let them out to void, and they would run satisfying those needs until they were ready. Then I would sit on the ground and they would love all over me. I liked deer hunting, but I especially enjoyed that moment with the dogs. Most often, they smelled bad, but so did I, so we accepted each other. We were all pot lickers.

What is a "pot licker?" I don't know if there is a formal definition but I propose one: A pot licker is a creature of perhaps sound ancestry who loves hunting with a passion, who often looks questionable, and who never turns down a meal. Often ribs show plainly as they run their hearts out and simply cannot eat enough to keep on weight. That whole pack, including me, met that definition, so "pot lickers" we were.

I learned so many lessons from that pack, and most involved what you can accomplish if you NEVER give up and that sheer joy and drive can overcome pain and even extend life. They grew older but not old, until the end was very near. They didn't get mean and crotchety as they got older. They taught me that activity and heart can forestall growing old and postpone death, that the end should come before the stretch, and that you can stay young and love hard right to the end. We don't have to grow old and crotchety.

Their drive was unbelievable as I saw them accomplish things that we humans would think impossible. Some were so beaten up, torn up, and ugly they should have been dead a long time ago. Yet, they ran right through pain until the run suddenly quit forever. As Guy Clark sang, "They didn't know they couldn't fly so they did."

Now, many years later with the light faintly glowing at the other end of my tunnel, I am trying to put those lessons to life. I refuse to quit, I want to be part of the pack until the end, most importantly I will not be crotchety, and I will love. I hope they are as proud of me today as I was of them then. To my family: When you bury me, put a 6-inch cap of concrete on top, elsewise when the next hunting/fishing season opens, I'll be back.

The Yellow Jeep—
One Sexy Beast

She was an old World War II surplus Jeep with a cloth covering that could only jokingly be called a top and two doors (also cloth) that could be added in really cold weather. Most of the time, however, neither the top nor the doors were on, the front window was hinged down and latched onto the frame. There was very little between the passengers and the outside world. She was raw-boned, dented and rusty, but she was beautiful in my eyes, as my dad had decided that she was a good first vehicle, so I was told to drive it until I earned the right to a classier vehicle (which turned out to be a really sexy '51 Ford—another story).

When I first got her, she was red (sort of) but it was difficult to determine what was the red paint and what was rust. I figured red wasn't appropriate, so I bought several cans of yellow paint and a large paint brush. When I was through, she was a gorgeous canary yellow jeep with lots of brush marks adding to her distinctive look. I guess we were the envy of our small town, West Point, Virginia, for lots of people would look and shake their heads as we'd drive proudly by on our way to the end of town, turn around and drive to the other end, turn around and drive back to the first end, etc. ad infinitum. Since we lived in a small town, the whole trip from one end to

the other took maybe 10 minutes. Periodically, we'd stop at Broadus's Dairy Queen to get an ice cream cone and go back to the driving. Of course, there'd be lots of other young boys riding with me, all in search of young girls, who never seemed as anxious to ride as the guys did.

My favorite passenger was an English Pointer (bird dog), Spec, who was always willing to ride as long as he could sit in the front passenger seat. He would very patiently sit there knowing that when I got an ice cream cone, he would also get one. The cone would sit nicely in a cup holder and he'd lick as we drove along. When he licked down to the bottom, he'd look at me with a "well, do it" expression. I'd reach over, pick up the empty cone and hand it to him. Then, and only then, he'd chew and swallow the cone. We were quite a team.

One day Bob Jackson and I went fishing at "the pond." After a good day of fishing, we were driving home on the back road near Five Mile Hill, and I decided to see just how fast the Jeep would go. With a downhill start, my foot to the floor, and "double clutching" through the gears, I could get her to 45 MPH if the wind was in the right direction. We never once got a speeding ticket.

Those were good days. I was the wealthiest kid around with a Dad who completely understood me. We didn't have much money, but we sure had fun.

I only had her a couple years before Dad decided I was old enough to have my own real car. Knowing that my pursuits might be more successful in a car with a back seat not cluttered with fishing poles, I dove into the next car with great gusto. Those other pursuits were no more successful, but they sure were fun. I'll tell you about that later.

NO GOT
A Fond Memory

He would show up every Tuesday and Thursday, mid-morning. We kids would be playing in our yards watching diligently up Lee Street in West Point, VA, for his horse-drawn cart, up by Eddie's house a full quarter mile away. The race would start as we ran to greet our friend "No Got" and his horse-drawn wagon. He would see us coming and clean a spot amongst the vegetables for us to sit. "Hi, No Got!" would be our greeting, he would smile and continue walking as we ran around and jumped on the back. Once there, we knew not to move much or our friend would give us a stare that could melt an iceberg. You see, No Got knew exactly two words of English (as far as we knew)—"no" and "got."

The wagon was a non-descript four-wheeled one that had seen its better day but was still usable; the horse was an old gelding or mare (let's assume gelding) that also recognized us. He never put an ear back, shuffled, or anything that showed alarm. His job was to plod very slowly down the street on a memorized path covering all the homes on Lee Street, for in that wagon were vegetables that No Got was trying to sell.

The horse's hooves would go clop… clop… clop in an extremely slow rhythm. That, accompanied by the laughter of children, would be the only sounds as this entourage slowly moved down the street. The pace was so slow that No Got could walk house-to-house on both sides of the street. The lady of the home would tell him (how I do not know since he could not speak English) what she wanted, he'd walk back to the horse, ground tie him, put the order in a paper bag, take it to the house, and take the money. I never saw him make change so they must have given him what they thought was right, and he accepted it. He'd walk back, and through some signal, the horse would start again clop… clop… clop on down the street. No Got would continue walking.

If the lady wanted nothing, he'd walk to the next house, crossing the street as needed, and the horse would exactly match that pace. I never saw the horse get ahead of him, and had it done so, I'm sure there would have been communication to the contrary. If the lady wanted something he did not have, his only response would be "no got," hence his name.

About one quarter mile past our house, we'd hop off, say, "Bye, No Got!" and walk back home. It's important to note that in that one half mile together we may have seen one car, for we all were wealthy paper mill working families with very little money.

Over 60 years have passed, but the memory is there. He was a Polish emigrant whose family came to work in the paper mill, and this was his way of helping financially. I know my parents made it a point to buy, even when they really did not need anything. You see, we had a garden of our own, but that was unimportant.

A Deer Huntin' Christmas

We could see our breath as it was a chilly Christmas Eve, and the Turner men were going deer hunting as we did for years. Percy (Dad), Ken (elder brother) and I (kid brother) hopped in the old red Jeep and drove north. The heater was putting out as much as it could, but with the top and sides of the jeep in their sad state, it was COLD. Shotguns were cased in the back seat with me. Dad was shooting a semi-auto Browning, Ken was shooting a model 12 pump, and I was shooting a single-shot Savage, all 12 ga.

Arriving at the meeting site, men were standing around talking, dogs were relieving themselves, and all were excited to have one more hunt before Christmas Day. Christmas Eve's was a short hunt—all wanted to go home early to be with family—so we hunted the "triangle," a small, maybe 300-acre, wood lot surrounded by dirt roads on three sides and the rough shape of a triangle.

Dogs were turned loose, men were in the woods at likely ambush spots, and all were in a merry mood, except maybe the deer who were bedded. There, a cold trail bark started the day. The bark became hotter as the dogs got closer, and finally, the deer was up and running. Since the woods were small, the hunt was over fairly quickly, and a buck was down. A venison roast was assured for most of the hunting party, and it was still early enough to do it again. By 2 p.m., another deer was down, assuring all a roast for Christmas. Back to the skinning tree the party went. Guns were put up carefully, and the skinning started. Yes, some watery-looking clear stuff in a jar was passed around, but the Turner family did not participate, and the guns were up. (Later in life, I did participate in that part of the ritual, but back then, we passed.) The meat was divided among the party, and we all want home.

Walking into the house, the smell of the cedar Christmas tree which Dad "shot down" (honestly!) was a comforting aroma that hugged me like my momma did. No fake Christmas tree for us; it had to be a red cedar. There on the fireplace mantle was running cedar greenery with holly spread throughout for color. (Writing this, I can smell those smells today. I loved it.) On the front door was a homemade Christmas wreath with pine cones, more running cedar, and holly.

That night, we each got to choose one Christmas present to open. Often the gift was a gun-cleaning kit, shotgun shells, fishing tackle or other very useful outdoor tool. Never was it expensive, as we could not afford much, but always it was full of love. "A man's wealth is evidenced by the love and laughter around his campfire," some wise philosopher said.

We were in bed by 10 o'clock, for Christmas day would come early. Mom would be up first, adding to the smells mentioned above with fresh-brewed coffee and frying country ham. Then we were off to the Christmas tree, with coffee for all except me—I had hot chocolate. Christmas paper flew everywhere as we opened our gifts and wished each other, "Merry Christmas!"

That year, I got a warm, fluffy, pointer puppy—again, something useful for what our family did. (I have written about Spec before; he turned into a good hunting dog and a wonderful best friend.)

For Christmas dinner there was a large turkey with homemade dressing, mashed potatoes, creamed green beans, a few slices of venison roast and copious amount of desserts. My favorite was pecan pie with vanilla ice cream.

After dinner, we sat around awhile trying to get to where we could move. We helped Mom clean up some, but not as much as we should have. As if on signal, Dad stood up and said, "Let's go look for dogs." We had about 3 hours of daylight left, so we hopped in the Jeep and rode out into the country. This was my favorite part of Christmas.

Riding around, we'd stop periodically to listen for running or trailing dogs, and we'd talk a lot. Not seriously hunting, we shared stories, fears, problems, and while not solving all, many problems seemed to just disappear. Oh yes, the shot guns were cased in the back of the jeep, but of course, we could uncase them if we had need. Going home about dark, we sat by the fireplace until late, then went to bed.

Abundant amounts of leftover turkey and venison were consumed in the next few days along with what desserts might have survived. Thanks Mom, Dad, Ken, and Spec.

Memories of
Early Hunts with Dad

Wayne Turner

Decoys bobbing in the night's pale moonlight,
Wings whistling, ducks flying over, as day slowly steals from night
The warmth of a father snuggled close to me
    As a flock, wings cupped, pitched in from lee,
A gaggle of geese, whose sound reaches from afar
    As we pour coveted hot chocolate from a jar.

    THESE ARE THE MEMORIES OF EARLY HUNTS WITH DAD

The smell and weight of marsh and salt air
    As we sneak, hoping to surprise them in their lair.
The wind so bitter, the cold so deep,
    Yet, not once uttering a complaining peep
For the flock is circling, making its third pass
    Warming my wet and very cold ass.
The thrill of that first downed drake wood duck
    Who fell at my shot, so charged with hope and luck.

    THESE ARE THE MEMORIES OF EARLY HUNTS WITH DAD

The sight of those two magnificent mature bucks
    Crossing the field, as we sat watching, in the truck.
The train whistle sounding from far up the rail track
    Prompting Dad to say "That means it'll snow before we're back".
The sounds of Walkers, Redbones, and Blueticks bellowing in the dim
wood's light
    As we wait for the excited bark of running by sight.
The foot race, deer and I, to the crossing lane
    Learning, it's not the deer I hunt, it's Wayne.

    THESE ARE THE MEMORIES OF EARLY HUNTS WITH DAD

The sight of the buck swimming a river to hide in the other-side marsh
    As Dad and I watch from a cliff, the cold and wind so harsh.
The men, who to the shore they came,
Hoping those antlers to lay claim.
The buck, in clear view, yet motionless, seemingly not in sight
    As they left, he swam back overcoming their armed might.
Dad, in a rare moment, saying, "Let's leave him be;
    A buck like that should tomorrow see."

THESE ARE THE MEMORIES OF EARLY HUNTS WITH DAD

Why then, did not the pain of fatigue, cold, and wet
    Overshadow the pleasures of this early age rite?

Because my Dad and I shared an unbelievable love for each other.

Thanks, Dad.
    Yes, we truly did pour hot chocolate into a jar, wrap it in a towel, and take it with us. I don't know why we didn't use a thermos; perhaps we couldn't afford one.

Kathy—
My Love and Fishing Partner

She had the longest, prettiest legs I ever saw, when she met me in the hall of the Radford University dorm (Virginia) for our blind date. I almost swallowed my tongue, but I managed to keep my cool, I think. I had just been fishing, so there was a small bait bucket of hellgrammites (great fish-bait larvae with long ugly pinchers) sitting in the car. While driving to meet her, the bucket turned over, and hellgrammites started crawling around inside the car. They were valuable bait, so I pulled over, caught them again, and put all of them back in the bucket (or so I thought). Later we were driving to a party and carrying on quite a conversation when I heard her say something like "(censored), what the heck is that?" A hellgrammite was crawling up her seat. She didn't scream; she looked intrigued, and I thought, "Wow, this is cool." I caught it and a of couple others I had missed, explaining they were the larvae stage of a fly and great fish bait. She asked if she could fish with me someday, which we did and still do almost 50 years later. By the way, our second date (or so) was to catch hellgrammites and crawdads in a stream just outside Blacksburg, VA.

Kathy appears in most of the stories from here on, so I won't repeat a bunch of them, but a few might help set the stage. She was younger than I, and as a friend recently said, "you married up." We raised some young 'uns, a bunch of bird dogs, a few barn cats, some raccoons, and various and sundry other animals. You'll meet all of them in the stories to follow.

I remember well the day I fell in love. We were fishing with some friends; the catching was poor so we took to drinking a beer or two and shooting the bull. I heard her say, "Would you like some of this?" I turned, and she had some crackers laid on those legs (did I say she has pretty legs?). She was putting sardines, catsup, and onions on those crackers and was offering me some. One thing I loved then and still do was sardines with crackers, onions, and catsup. I decided she fishes, is beautiful, fixes sardines and crackers, and I like to be with her. That did it. All I had to do was convince her, which I did.

I was in graduate school when we met, and I lived with Bob, a ne'er-do-well that I loved like a brother. Bob and I would start each semester by buying a great bottle of bourbon and a cheap bottle. The great stuff would

go into a large flask, and the cheap stuff would go into the "good" bottle, to share with our dates. On about the second or third date, I poured Kathy a glass of the cheap stuff from the good whiskey's bottle. She took one sip and said, "That's not Wild Turkey." Once again, I was impressed. We don't drink much bourbon anymore, but we do drink wine, and her taste buds haven't deteriorated. I buy her good wine and always will.

As you will find out, Kathy casts a beautiful loop with her fly rod and is very meticulous in her approach to fishing a hole. She starts at the lower end and very slowly works her way through the pool, giving every trout in that hole a sore mouth before moving on to the next. I have never seen anyone fish so carefully. In the meantime, I have covered about a half mile of stream catching many fish, but missing many also. We make a good team.

We dated about 2 years before I proposed. She said yes, and we settled down in Blacksburg while I finished my PhD at Virginia Tech. Four years after that, we moved to Stillwater, OK, where I was a professor of industrial engineering. Most of the stories to follow occurred in one of those spots. Here is one about night fishing that involves her.

Big Black Jitterbugs on Dark Nights For Large Black Bass

"Glug, glug, glug," the Jitterbug sounded as I cranked the reel of the level wind, but I couldn't see a darned thing. I was on a local farm pond that I knew well, so I knew there were no trees, but I had no idea what might have floated in. "Glug, glug" the lure methodically worked its way slowly to me as I didn't want it to go too fast. "Keersplash!" A bass went for the lure, but she missed. I stopped my retrieve very briefly and started again, "glug, glug." "Keersplash!" This time the take was solid as I felt resistance. I had chosen a moonless night, so all I could see was black ink. I'm not sure what she saw, but I think she struck at the sound. Unable to see my prey, it was hard to resist, but I didn't set the hook until I felt the fish, as I have found the hook-up is usually solid if I just continue reeling slowly. If I had set the hook prematurely, the bass

TRYING THIS TECHNIQUE

I believe that darker nights are better and that black top water lures work best. As you know, I have a fondness for large black Jitterbugs (about double size seems to work well). Constant disturbance seems to be important, like a Jitterbug or crazy crawler would make. This seems to allow the bass to focus on location and strike at sound. Intermittent sound doesn't seem to work as well. I don't remember ever using crazy crawlers, but I'll bet your truck they work. Hula poppers or Pop Rs should work also, but the constant disturbance would require some attention on your part.

Of course, to get the magic, they must be black to match the dark night. Good fishing, and if you do well, thank my wife and my dad.

would have lost the lure because she couldn't see well either.

I set the hook, and it was a nice one. I had no idea what direction to try to move her, so I horsed her as much as I dared. Smaller ones, I water-skied across the top, but bigger ones like this took finesse. I used 14 lb test mono so I could horse her pretty good, and I have used as heavy as 20 lb test since the fish can't see well either. Anyway, soon we made our acquaintances. She was a four pounder, well-proportioned and fed—a real beauty. I kissed her and let the dog lick her one time before releasing. I had caught several nice bass in the same spot before so I continued to work that spot before moving down a few yards. I did carry a flashlight which I carefully kept off the water so I could move down the shore line.

Such is nighttime bass fishing with large black lures.

This story occurred about 45 years ago, and in the truck waiting for me were the two people I loved the most. My girl, from Washington, DC, whom I was then dating and to whom I have now been married 46 years was in the truck with my dad who is responsible for my doing such ridiculous things. He started all this, so don't blame me. After this night which happened on a local farm pond, they both sometimes joined me as they became believers. Dad is gone, leaving fantastic memories, but Kathy and I continue to fish together often.

As an aside, they had the windows down, listening to the frogs croaking, when they heard the bass hit. Then I heard two truck doors slam and footsteps as they watched me release the behemoth. They both tried and caught fish that night on that same large Jitterbug.

Cold Weather Bass

While finishing my PhD at Va. Tech, Kathy and I lived on a mountain lake that was full of largemouth and smallmouth. I really enjoyed chasing them in warm or cold weather. We didn't have the pot or the window, so all I had was a 12-ft. Jon boat with a small electric motor. We'd cruise across the lake at one or two knots.

First, I had a major secret (this was 1969 before jig and pig was popular) in that I'd take a heavy black jig (3/8 or even ½ oz) to which I'd tie a black pork lizard. It had to be a lizard not an eel or I wouldn't use it. Then, I'd surgically enlarge the split tail and "tune it." The lake was crystal clear, so I could make a short cast and watch it fall. If the surgery was right the jig would steadily fall but flutter all the way down. It drove the large and small mouth crazy. Most of the time I'd know a fish had hit as the jig would just stop falling or I'd feel a very slight twitch at which I'd set the hook. Oh, I used 4- or 6-lb test to get a smoother fall so the set had to be light, and I was good at it. If it reached bottom, I'd lift the rod and drop/reel, keeping a tight enough line to feel the strike. If the water was cold, I'd very slowly work the bottom. Many brownies and blacks and a few walleye fell to that rig but I only showed that trick to very close friends (probably fewer than five) with a life threat if they spread it. I lived there 4 years, and it was just starting to catch on when I left.

Most of my fellow PhD candidates were sticks in the mud about fishing, but there was one other (John) who like me would rather fish than eat, so we would go out often. One very cold day in the office, we both had cabin fever and decided to go fishing with our secret rig. It was around 10°F, so we'd have ice on the guides and we looked like the Pillsbury Dough Boy. I never drink while fishing so we had coffee, but when we got back…

Fishing was very slow literally and figuratively, but we solved all the world's problems by talking. One cast I hooked bottom and started to reel in, the log starting up slowly. I'd lift, reel as I dropped the rod, etc. until the log materialized into a huge largemouth on the surface. Start the Keystone Cop routine! John and I tried to find the landing net. Making a couple of swipes, we found the net was too small, so I grabbed her mouth and lifted her aboard with that jig and pig hanging out. We raced back to my cabin at one or two knots where we weighed her at between 6 and 7 lbs (I really don't remember the details). In those days, I didn't have a lot of equipment, so my fish scale was the bathroom scale, with and without my holding the bass. That was a great day, and of course, I have broken that record several times since, but none has been bigger.

Jon Boats & Jitterbugs

After Kathy and I were married and about 10 years later, I was fishing with her brother who was as big a fishing nut as I. Tommy said he knew the lake, and I shouldn't worry about not being able to see. Yeah sure.

We launched the little Jon Boat in total darkness with only a small flashlight to help. We carefully tied on black lures and yes, of course, I was using that big black Jitterbug.

I flipped the bail on the Mitchell and let line peel out with the weight of that big Jitterbug pulling it back. Here we go, glug, glug, glug… The noise was somehow mesmerizing, allowing me to concentrate totally on the sound. The boat glided along under the power of that little Sears electric motor, and we were fishing. Tommy apparently did know the lake for we did not run into anything other than a few limbs.

"Keersplash!" A bass was after one of our lures. Of course we had no idea which one, so we just hung on and continue moving. "Keersplash!" She tried again. "Keersplash!" And I felt a heavy fish. The rod and I went to work. A few minutes later, a 3-pound largemouth was laying alongside the boat with that heavy Jitterbug hanging out the side. I removed the hooks with my pliers and released her. Next time, she should be larger. In those days, we did keep some bass, but always smaller ones in the 1- to 2-pound range. Larger ones deserved to go back and make millions of smaller ones.

Tommy and I continued catching fish until about midnight when we quit because we were tired, but they were still biting. We must have caught 15 large bass, almost all of which went back to the water; but a few smaller ones did go home for the table.

Late-night fishing continued for many years as that Jitterbug and I were best friends. Then one night, I cast him and "thud," he hit a large stump. With no boat and deep water, I tried and tried to get him back, but the line finally broke (of course, not the knot). I came back the next day with a float tube and tried to find the lure but never did. That was about 30 years ago.

In another 5 years, we had moved to Stillwater, OK, where we were raising two boys (Travis and Drew), horses, and bird dogs. On our farm, we had three ponds, two of which were full of hungry largemouth and catfish. One night, we took our fishing rods and went to the lower pond to try the black Jitterbug of which I had such fond memories. I had shopped everywhere and finally found some regular-size black Jitterbugs which we tied to our line. Yes, it was still good, as we caught a few bass but no huge ones, and somehow the magic just wasn't the same without that large black Jitterbug making the glug, glug noises. The combination of two young boys, a dad (me), a couple of bird dogs following us, and black Jitterbugs disturbing the water was still special, but I really missed the big black Jitterbug.

PART II

Raising the Boys
In the Great Outdoors

In 1974, Kathy and I loaded the Buick with all sorts of hunting-fishing equipment and two boys in diapers, Travis and Drew, to begin our move to Stillwater, OK, where I was to be a professor at Oklahoma State University. What a fantastic place to hunt, fish, and raise a family. We stayed there 33 years, hunted and fished all over the state, and truly enjoyed the experiences.

I immediately started turkey hunting and became rabid with turkey hunting fever. I had never been successful in Virginia. Since then I have harvested one to three turkeys each and every year including Eastern, Rio, and Merriam subspecies, and the boys have also hunted and killed many turkeys, so many of the stories are about turkey hunting.

We also quickly began quail hunting which necessitated bird dogs. Kathy, the boys, and I have raised and trained six bird dogs of different breeds, and they have been an important part of our family with Okie being the first. A Weimeraner, she officially raised the boys and showed them how to hunt. We now have one dog, a great yellow lab named Annie. In between were four other dogs of which you will hear stories.

In the 70s, 80s, and 90s, we hunted hard and well, averaging about seven coveys a day. Today, if we were to try, we'd probably average one unless we went to western Oklahoma or Texas. That my friends is a sad commentary on what has happened, as quail have all but disappeared from most of the quail country that I've hunted throughout the USA. Why? This has been discussed in many research papers, and there is a lot of research underway. Most agree the decrease in numbers is due to much habitat degradation. As a kid and as an adult, I could walk fence rows that had green briar, multi-floral roses, and other fine quail habitat bushes or "weeds," and I jumped (discovered and put to flight) many coveys. Now, there are very few fence rows, as farmers plow fence-to-fence. Good habitat in spots helps, but quail like to move around, so we need *large expanses of contiguous quality quail habitat to restore the numbers.* In very recent years, I have seen improvements, so hopefully the numbers will return for my grandson.

Happily, many other species have thrived. Many more turkeys and deer roam the prairies of Oklahoma today, perhaps more than ever. I researched it before moving. Oklahoma harvested 7,000 deer in 1973, and today the harvest is frequently over 100,000. The pursuit never ends, but the pursued do change.

Here are stories about all of us during the period between 1974 and 2000. After that, the boys had graduated from college and moved on with their lives. To this day, we still hunt together, just not as often.

Please enjoy,
Wayne Turner

Dwain Bland, Me, and
A Bunch of Turkeys

When Kathy, the boys and I moved to Oklahoma in August of 1974, I was an experienced hunter/fisherman but burning deep inside me was a desire to become a turkey hunter, or rather a turkey killer—I had been hunting them for a couple years while I was working on my graduate degrees at VPI&SU, Blacksburg, Va.

There was one old bird east of Blacksburg that lived high on a mountain. He was warier than I, but we had come to know each other very well. With an old shotgun that had belonged to my dad, and a Lynch Box Call that Dad had given me, I climbed that mountain many mornings to renew acquaintance with the old bird. He would always answer me but would hang up every time. In those days, we were taught to yelp no more than three times and repeat that no more than three times (nine yelps total). Today, I carry about four different calls each time and am getting the entire vocabulary in hand. Finally, the year before I was leaving, I was sitting along a trail calling when I heard a loud "whoosh" that sounded to this old swamp duck hunter like a huge flock of teal coming in. I looked up, and there he was.

Tiring of our encounters, he had decided to fly down the mountain straight to me. He passed overhead so close that I could have jumped and grabbed that rope of a beard. Sitting there with my mouth wide open, forgetting that I had a shotgun, I simply looked. He sat down behind a hill about 20 yards away, gobbled once, and walked off. With figurative turkey poop all over me, I chuckled a bit, said something about the bird's ancestry, grabbed my gun and got infected with a turkey-hunting fever that I suffer from today, 40 years later.

After settling in Oklahoma, I was reading a hunting magazine and ran across an article on turkey hunting by Mr. Dwain Bland of Enid, Oklahoma, about 40 miles from my home. I wrote a letter addressed to Mr. Dwain Bland, Enid, Oklahoma, and he got it! (Try that today.) In the letter, I told the story above and basically said "HELP." He wrote back and said, "Sure, how about next weekend?" So I grabbed the same shotgun, Lynch call, and my camo gear and drove to Enid.

We immediately drove to Vici, Oklahoma, and camped for the weekend. After setting up camp, he and I went to a shinnery (a dense growth of small scrub oak) and glassed the drainage below. Now, if you don't know western Oklahoma, you can't appreciate that it is almost a desert with a few elm and cottonwood trees in the bottoms and shinnery oaks on the hills. I said to Dwain, "Turkeys can't live in this." He said in return, "Wait."

Just about sunset, the bottom came alive with deer and turkeys. I still hunt the area, and folks, it is still a wildlife Mecca. We glassed and picked out a nice tom which we first tried to call. He showed no interest other than a gobble over his shoulder. When he went behind a knoll, we hauled donkey down the other side to get ahead of him. One half mile later, we sat up and called again. This time he double gobbled, but he still had an agenda that didn't include us. He was about 100 yards out, going behind a bunch of cottonwoods in the other direction. Dwain said, "Stay here" and flat-out ran around the same cottonwood trees in the opposite direction. That turkey saw this camo idiot running straight toward him and turned to run straight toward me. At 20 yards, a load of number fives ended his run, and I had my first turkey, further contaminating me with the fever.

Around the campfire that night, Dwain brought out a bottle of "Old Rot Gut," and we talked hunting/fishing until late at night. The whole time, I fondled that bird, looking at his beard and spurs as if they were going to disappear. That experience and

Dwain talked me into joining a group of turkey hunters in South Carolina that was starting a turkey hunting and conservation group. I joined in the mid to late 70s and have been a member ever since. Today, that is the National Wild Turkey Federation (NWTF), and they are a great group. If you aren't a member, perhaps you should check them out.

Dwain wrote several articles on turkey hunting and a book, *Turkey Hunting Digest*, DBI books, Northbrook, Illinois 60062. I'll bet you can still buy a copy from Amazon or some other source, but you can't have mine or the two I have saved for our boys.

Toward the end of our sojourns together, Dwain bought an old Damascus barrel front-end loader 12 ga. shotgun that became his weapon of choice. As I remember, he loaded it with about 60 gr. of true black powder and an equal volume of #5 shot which was lethal only to about 20 yards. That proves what our son, Drew, says today, "Turkey hunting ain't about killing turkeys."

night started a great friendship that lasted about 25 years before some unimportant event or two (like work) put a wedge between us. He called in birds for both our boys, and we chased birds all over Oklahoma and Arkansas. He was a great friend.

Believe it or not, I finally got to where I could call them myself, but somehow no bird was ever bigger than that 2-year-old bird with a 6-inch beard and ¾-inch spurs. (Last year, I called in an Eastern with 1-3/8-inch hook spurs and almost 12-inch beard, but it wasn't as big emotionally). I estimate that I have easily killed close to 70 birds since then, including Merriams, Rios, and Easterns, but I still suffer from turkey fever. Maybe just one more will cure me? I hope not.

Thanks, Dwain

May all turkey hunters know the thrill of long beards, hook spurs, and responsive birds.

Dwain and son Travis with a bird Dwain called in for Travis

A Cow, an Aggressive Bird Dog, Mike, and a Bellowing Calf

No animal was hurt in this episode as the cow and calf reunited at the end and everyone (except Mike who was bruised and battered) was happy. I laughed so hard for years that tears would flow when I recreated the scene. Want to hear the story?

As normal, when quail hunting with Mike, we were standing in the pasture at first light drinking coffee, waiting for a covey wake-up call. There, in the northeast corner, a covey was whistling to be sure all were present before starting to feed. By the time they grouped, we were into them and started our day with four birds before leaving the singles alone and going elsewhere. Mike was both the most serious bird hunter I ever met and a conservationist. We never took more than half a covey before leaving the rest.

Mike was a very close friend who bird hunted with me and our dogs for probably 10 years before unimportant events separated us and we lost contact. He loved bird hunting with a passion, and he hunted with vigor and intensity as you shall soon see.

Once, an Oklahoma cold front set in, and we had snow on the ground for about 30 days. Quail can handle snow for awhile, but Oklahoma snow turns crusty as it melts and refreezes, and the quail have trouble scratching for feed. We found our first covey, and the birds flushed and flew lethargically only about 20 yards before setting down and WALKING. I turned to say something to Mike, and he was already unloading his shotgun for the last time that season. Mike loved quail hunting, bird dogs, and quail, but in a fair-chase situation, don't get between Mike and his quail.

The opening paragraph took place on a friend's ranch near Stigler, Oklahoma, around 1982. Mike was hunting three Brittany Spaniels; I was hunting an English Pointer named "Sooner" and a Weimeraner named "Okie." (I never was very creative with dog names). Sooner had the best nose of any dog I ever hunted, and Okie was my best-friend dog as well as a single-finding expert. She would range close, and I honestly don't think I ever flushed a close single that she hadn't pointed first. I was and still am very fortunate in the bird dogs that adopt me.

We proceeded to have a great day until lunch. I don't remember how many birds we had, but we were well on the way to our limits, as in those days, we would find seven to ten coveys each day. After a peanut butter sandwich, a candy bar, and a drink of water, we soon were ready for the afternoon hunt. Mike believed that time was wasted if you didn't have dogs hunting in front of you, so off we'd go.

Shortly thereafter, Mike and I had separated, and I could hear him yell, but I could not see him. I heard a loud panicked "NO!" come from Mike, and a ruckus was shaking the bushes. Then, I heard some words that can't be put into print, so I called my dogs to heel. They were obediently standing by me when out of the bushes came the following scene.

First out was a panic-stricken cow with eyes the size of baseballs. She was grunting a bit, but pretty quiet other than that. Hanging onto the cow's tail was Mike's big male Brittany, and he was chomped down tight. Next on the scene was Mike who had somehow gotten a hold on the dog's tail, but Mike was very loudly yelling at the dog and asking me to help. (If you're keeping count, that's a cow, a dog, and an angry hunter chained together). Mike couldn't keep up with the dog and cow, so he sort of skied behind the dog, busting brush in the process. Finally, a calf that saw his mother running away was beside Mike bellowing.

Immediately, I did what any red-blooded hunter would do and started laughing so hard that I couldn't have helped him anyway. Mike saw this and yelled, "It's not funny," bringing up a reference to my ancestry, but again, I can't put that in print. I replied between laughs, "Oh, yeah it is!" My dogs didn't understand, so they broke and ran alongside watching. Finally, I grabbed Mike's dog, so Mike and I were both hanging on. The dog let go of the cow, the cow and calf had a reunion, and they walked off peacefully. No one was injured except Mike.

Mike put his dog on leash until we got well away from the cow and calf before releasing him. We continued our hunt in silence as Mike was not talking to me, and I was splitting a gut trying not to laugh, for you see the only critter hurt was Mike who was limping and scratched from the briars. We got our limit, went to the ranch house, told the rancher, and went to the motel where we had a drink. Today is more than 30 years later and the first time I have put this in print. I still bird hunt and always will.

Thanks Mike, Okie (I will always miss you), and Sooner.

Take Your Children Hunting; You'll Both Always Remember It

Oklahoma is truly blessed with significant amounts of great public hunting areas in terrain varying from beautiful cypress swamps (SE), through pine- and hardwood-covered mountains (SE), through prairies covered in oak and tall grass (central), all the way to high-plateau dessert (NW). There are public hunting areas in all of these. Remembering his youth there, my younger son, Drew, recently made the statement, "I can't imagine having a better childhood." I thanked him for sharing his childhood with me. This is the story of Drew's first deer in NE Oklahoma, right on the Kansas border.

Big bluestem, so tall that people on horseback disappeared; elk and grizzly bear roaming the plains; and greater prairie chicken so thick and trusting that settlers could easily bring home a dinner all describe what the plains used to be. In northern Oklahoma's Osage Hills, all this used to exist in plenty and still does in spots today. It is the home of perhaps the nation's most fiercely independent ranchers who are also among the nation's most conservation-minded landowners. The area is a truly rich environment in history, flora and fauna, and is where we used to own a small ranch that is now a public hunting area. That ranch is where this story took place, on a hill overlooking a valley and small town in Kansas called Elgin.

The grasses were still rich in big and little bluestem, and the wildlife was still exciting, but because of people and fire suppression, trees had invaded. First, the prolific but scrawny blackjack oaks moved in. Scarcely ever more than 12 to 15 feet tall, they tried to grow as thick as hair on a dog's back. The good news is that fire thinned or even eliminated them, and while alive they scattered large amounts of small, bitter, but protein-rich acorns. On thicker and richer soils, post oak trees took hold, and once they reached a certain height, they tolerated grass fires. In the bottoms were sweeter acorn producing white oaks, some of which were older than this county, and streams that were crystal clear with Kentucky and largemouth bass as well as flathead catfish and the promiscuous bream/bluegill, all waiting to be caught. Oh, and the hills were alive with whitetail deer and the sounds of turkeys. Of course, all this was why we bought the ranch and why we sold it to the state for a wildlife area when we left to follow a dream to Montana. The fishing is another story that we shall share later; today, let's get Drew his first deer.

Drew was 12 years old hunting with my old pre-64 Model 94 30/30 lever action. That old rifle was dear to me, so it is fitting that he hunted with it. The rifle also blended in well with the cowboy history of this area, as there were ranches that still worked cattle on horseback, but most had gone to 4-wheelers. Drew had taken a couple of hunter safety courses and had been raised around firearms his whole life. He was extremely safe and careful. He won an award for making a presentation on firearm safety to the local 4-H group when he was 10.

Our camp for these hunts was a large wall tent with a wood stove and cots. That bucolic scene is a vital part of the lore and memories we have, for as the boys were growing, we spent many nights in that wall tent. The stove always leaked a little, filling the tent with a pleasant aroma; but, we were always careful to have adequate ventilation through the tent. On cold evenings, we'd come back from the hunt, stoke the stove, and warm ourselves with the emanating heat. On warmer evenings, a glowing campfire was all that was needed. Our campfire pit, ring, and stack of firewood were always a welcome sight when we walked back to camp cold and tired. Many stories were told and retold in the evenings around that fire and inside by the stove. Remind me sometime to tell you about the big charging boar hog and black

This hunt occurred on a track of land we once owned right on the Kansas border. It is now an Oklahoma public hunting area that is blessed with large-bodied and racked deer, turkeys, quail, prairie chickens and many other types of wildlife. From the right spot on this land, the lights (both of them) of Elgin, Kansas, provided a nice view from the campfire. Hunt camp always consisted of our two sons (Travis and Drew), friends, and myself. Many memories and great stories occurred there. We moved to Montana and sold that land to the state so it would always be part of the hunting heritage of Oklahoma; we hope you enjoy it. That rifle is still in my gun safe and will be until the day I die or perhaps when Ethan, Drew's son, is ready. Wouldn't that be great? At that time, it was just a fine older rifle, but today, it is a collector's item.

Top: Drew, his first deer, and me.

Left: Travis, his first nice deer, Model 94 rifle and me.

powder kill; that story has been around for years and was frequently retold (sometimes, the story would remain the same each time, but often a few facts would change).

The morning of this hunt, Drew climbed a large post oak where we had created a comfortable sitting area on a tree stand and cleared shooting lanes through the branches. I held the rifle until he was settled; he lowered a rope, and I tied the unloaded 94 to the rope for him to haul it up the tree. He took the rope, tied himself in the tree (now we use a safety belt), and loaded the rifle. I waved goodbye and walked down a deer trail to my stand with more than a little apprehension. Today, 35 years later, we have moved our camp to a mountainside in Colorado, and I still have that apprehension when he takes off over the mountain after elk. He will always be my son, even though now he is a better outdoorsman than I, and both of us are considerably older.

My stand was

Elgin, Kansas, used to bill itself as "A Town Too Tough to Die" as shown on the sign over the main and only black-topped street. Margaret's Café was a priceless visit in history and where we often ate lunch.

Elgin was once a rail head at the end of a cattle trail. Cattle would be driven across Texas and Oklahoma to the rail head where the "longhorns" would have the horns chopped off so they could fit in a rail car. A friend of mine's dad was a "chopper." This fascinating history could be traced on the walls of Margaret's Café.

Walking into Margaret's we would be greeted by 15 or 20 ranchers and local oil field workers. I always had our two sons and often other boys with me. The first time or two, everyone would stop eating and look at us like "who are you?" Soon they greeted us as old friends and took the boys under their wings, which was really good for the boys. We loved it there.

The walls of that café were a history book with pictures starting in territorial days and proceeding to the time we were there. Some of the pictures took some explaining on my part; but I assure you the boys learned as much history from those walls as they did in class. By the way, Kathy and I had a philosophy: "Don't let school interfere with your child's education." The boys were frequently taken out of school to hunt with me, and they now have graduate degrees and great jobs. I think their education was complete.

With all my heart, I hope that Margaret's Café is still there and that those ranchers are greeting other young kids today. Our country badly needs more such places. Take your kids hunting and fishing.

½ mile from his at the base of a large post oak overlooking the intersection of a couple deer trails and an opening to the valley below. Daylight was just starting to win the battle with night, and the woods were coming alive—my favorite time of the day. I absorbed all this and even saw some deer in the distance; but this was Drew's hunt so I sat tight. At 10 AM (now we will often sit all through the day), I decided to give him a break and walked back to his tree. He saw me, kicked the shells out, put them in his pocket, tied the rifle to the rope through the lever and lowered it. We didn't talk, for the deer were probably still moving and it was rut so they might move all day.

Drew was lowering the rifle to my waist level where I was supposed to untie it, make sure it was unloaded and wave him on down. My peripheral vision picked up movement, so I slowly turned my head to see a nice young buck 75 yards out moving our way. Since this would be his first, we had agreed to take a nice yearling if the opportunity presented itself. Well, there it was. However, Drew did not see the deer and was busy lowering the rifle.

I whispered "Drew, Drew" several times until he finally looked my way (in the meantime, the buck was getting closer). I pointed at the deer and stammered "sshhoot him" (dads get more excited than the kids at this point). Drew started the rifle back up the tree by the rope until he had it in his hands but he couldn't shoot, since I had taught him to tie the lever with the rope; he silently untied that rope as I stopped breathing but continued to watch the deer get closer. Thank goodness for Oklahoma wind. Drew untied the rope, opened the lever, and put one shell in the chamber. The wind hid all that noise from the deer that was now so close I could hear his footfall and see his breath condense in the cold air. Drew shot the deer that was no more than 15 yards away, making a perfect double lung shot clipping the top of the heart. The deer died almost instantly.

Drew lowered the now empty rifle and tossed me the shell which I saved; then he climbed down the tree. We went to the deer where he checked for life by touching its eyeball with the rifle. We said thanks, field dressed the animal, and floated back to camp; both of us talking all the way.

I have often tried to describe the feeling I had at this point, and it's impossible. My heart was swollen with pride and excitement, but at the same time, we both felt remorse. After all, an animal just died to feed us; we owed him a lot. Some indigenous tribes believe that at this point the deer's spirit merges with the hunter and the hunter is a better, stronger man for it. I believe that, and I know Drew didn't stop talking about the kill for two days. That night around the campfire he shared the story with anyone who would listen, and each time he told it, I was a little prouder. Today, he could tell you the whole story just as I did and at least the details would be recognizably similar. After all, it's his story.

Thanks, Drew, and thank you for reading this.

(When you get too old to hunt, you teach others how to read scat, and you write about the old hunts.)

Downtown Elgin, Kansas, and Margaret's Cafe

The Great Bird Dog Capers

Adam
Eats the Truck

I found he was a great bird dog with an ability to find birds when others could not, but he also did a great job retrieving waterfowl. The only problem was he could not sit still in a duck blind, and at the first shot, he exploded out of the blind looking for birds. A training collar would not even slow him down, so I was at a loss as to how to use him for ducks.

No problem, I thought; I'll just leave him in the truck, and at the end of the day, I'll let him out to retrieve the ducks. On the first day, I let him out while I placed the decoys and he ran his ever constant sprint around the pond. Then, I walked him back to the truck and locked him inside where he had always been content, since it smelled of me. I believe he was OK with that, until I shot the first volley. Then he exploded into a whirling dervish inside my truck.

I came back, let him out and guess what, it worked. He ran to the pond and started retrieving ducks for me while I smugly watched. Then, we walked back to the truck, I started it and pulled on the seat

Adam

Adam was a German Shorthair that had two speeds, fast and faster. He dearly loved me, and he hated to be left behind. By the time he came along, I was an experienced dog handler, although you will never know it from these stories. Adam was certainly a challenge.

belt to buckle in. The seat belt came out in my hands. Adam had apparently decided the seat belts were holding him in the truck and had chewed every one (five total) into two pieces. No other damage, just the seat belts. About $500 later, I had a whole truck, and we tried again.

The next time, I put Adam in the back of the truck and closed the topper camper door. I even locked it in case he figured out that part. At the end of the hunt, I came back, lifted the topper door, and opened the tailgate which fell totally off the truck and landed on my foot. Thank goodness for thick waders, but it still hurt like crazy. Adam had decided the cables were holding him in, so he chewed through the wire cables holding the tailgate, until there was nothing holding it on but the latch. Several hundred dollars later, I had a whole truck again.

Adam simply wanted to be with me. By the way, I solved the problem by putting him in a crate in the back of the truck. By then he had figured he was going to get to retrieve anyway and quit fighting. Besides, his teeth were worn down considerably.

The Fishing Rod Terminator

This is another Adam story. In addition to enjoying bird hunting, he loved to ride with me in the front passenger seat, watching the world go by. He would object when Kathy told him to get in the back, but he would comply. I loved him enough that I let him ride there even though he left half his hair on the front seat when he got out.

Previously, he had always waited for me to say "kennel" or "get in," but this time he took it upon himself to launch. I was going fishing and had placed four expensive fishing rods and reels between the two seats as they were "safer" there than in the back. I placed the rods carefully between the seats, turned away and bent over to pick up the rest of the tackle. A flying German Shorthair sailed over my head into the front seat, as he wanted to go. He broke all four rods in two, settled into his seat and smugly looked at me as if to say, "Let's go." In case you're keeping a sum of the cost of owning Adam, we are now into several thousand dollars—well beyond what I paid for him.

Adam had an accident running into a parked tractor while chasing a retrieving dummy, and broke his leg in several places. He is now finding birds in heaven, leaving behind quite a set of memories.

Sooner

The next two stories involve Sooner, an English Pointer that had the best nose I have ever experienced before or since. I have seen him lock on an air scent 50 yards upwind, then slowly move in. He loved to bird hunt but he had to find birds. If you didn't find them in the first 3 or 4 hours, say goodbye—he was off to the races to find them for himself. No problem; I'd return to where we let him out, lay my coat on the ground, and come back early the next morning. He'd be lying on the coat looking at me like, "Where in the world have you been?"

The Bread Aisle Scent Marking

Sooner was difficult to keep penned. His pen was heavy-duty fencing, 6 feet high, with cattle panel for flooring and woven chicken wire for a roof. (Remember this chicken wire for the next story). On this occasion, Sooner GOT OUT through the chicken wire by chewing a hole in it. Once he obtained freedom, he went looking for birds or the kids. Our son Drew worked at a local grocery store, so Sooner trailed him to the front automatic door which slowed him down for a few seconds. Once he saw the door open, witnesses to the crime say he bolted through it and the chase began. Boys in white aprons, including Drew, chased Sooner through the store bolting by startled shoppers. On his way by the bread aisle, he decided to "mark the bread." (If you don't know what that means ask someone who owns a male dog.) Drew said he saw Sooner barely slow down to spray the entire row of packaged (thank God) bread. That slowing down was enough for our son to catch him and call me. I sheepishly drove down and took Sooner home.

The store manager was a bird hunter, so when I offered to pay for the bread, he laughed and said it was no problem. (You may not want to know the rest of the story.) When I asked what he did, he said, "I just wiped it off really good." How does your bread taste?

The Great Scrotum Hang-up

On another occasion, Sooner tried to get through the chicken wire, but I had woven the top with thick wire so he couldn't complete the hole. Once he got a little opening, he jumped on the roof of his dog house and tried to climb out the hole. Most of him made it through, but his scrotum hung up on the chicken wire, and that led to a problem. He was mostly out, but a vital part was still in the chicken wire. I guess he decided that getting out wasn't the most important thing in his life, so he climbed back through the hole and jumped down to the dog house; however, his scrotum did not make it that way either. At that point, he was standing on his dog house on tippy toes, to take some of the pressure off, waiting for help.

Kathy walked by a window and noticed Sooner standing still on the roof of the dog house. It was very unusual for Sooner to ever stand still, so she laughed and moved on. A few minutes later, he was still there so she went out to check on him. Seeing the situation, she quickly cut the wire and Sooner jumped down. He was sore for a few days, but all parts continued to work well (perhaps unfortunately). You know, he never tried that exit strategy again.

The Fishingest Dog

Sooner loved to fish. He would go to the little pasture pond with the boys to sit and watch. They were fishing with night crawlers as all young kids should do, so they had bobbers. Sooner knew that when the bobber went under there was a fish, so he would bark and run around which was helpful to two young boys with typical attention spans. However, Sooner was not patient, so if the bobber didn't go under quickly, he would start to tremble like a bird dog thinking the covey was about to flush. Sooner would plunge in, swim out to the bobber, carefully take it in his mouth and bring it back to the boys. We guessed he wanted fresh bait on the hook. We couldn't break him of this and frankly, never tried very hard.

Okie

The lore is that God gives us one really good bird dog in our lives. I disagree, in that I have had several, but if I had to choose one, it would be Okie. She was a Weimeraner that the family and I picked up one Sunday afternoon in southern Oklahoma. She came home with us at 12 weeks old and never once cried. Somehow she knew she was home. She loved to hunt, love on me, and protect the family.

At the time, we lived in copperhead (poisonous snake) country, and our sons were young. We let Okie out with the boys and she would protect them from all danger including snakes. She would corner the snake and bark until I came to the rescue. We loved that dog.

Okie Waiting by the Boat

The Gopher-Catching Team

We moved to Montana where there was snow cover much of the year, and this dog loved gophers. We also had a Yorkshire Terrier named Howdy who liked them just as much. The terrier weighed 3 pounds, and the Weimeraner weighed 50, but they were the best of pals. They would decide to go gopher hunting.

Walking through the pasture, the Weimeraner would break through the snow and the Yorkie would "float on top." Periodically, they would stop, Okie would tilt her head so her ear opened and listen. If nothing, she'd move on but if she heard something under the snow, she'd pounce and start digging. She would dig through the snow and into the dirt, back up, get out of the way and let the Yorkie take over. The Yorkie would then run down the hole looking for the gopher. If nothing, she'd back out and let Okie dig some more. Most of the time, they would find that gopher together, and they did not fight over it.

The Skunk "Hit Woman"

Okie hated skunks. Most of the time she'd give them room, but periodically, she would consummate the task, for as a pup, skunks kept stealing her food, so she learned to really dislike them. We'd let her out to walk with the boys to the school bus stop. Once the bus came she would walk home and wait until the bus showed up at the end of the day when she'd meet them again.

One day, she wasn't there, which was very unusual, so the boys were concerned. Walking toward home, they heard this loud "klunk, klunk, bang" coming out of a culvert. Bravely looking into the culvert, they saw Okie, skunk in mouth, swinging her head side to side hitting the culvert in each direction. Obviously, the skunk didn't especially like this situation and responded in the only way it knew.

Okie came limping home with eyes watering and smelling horrible. We tried all remedies and while some helped, only a lot of time cured the smell.

Here, if You Can't Hit Them—

On one hunt, my friend Steve and I went west of town to find some birds, and find them we did. We had the first covey point within 10 minutes and had a perfect opportunity which Steve and I proceeded to blow. We missed two shots each, laughed and started after the singles. Okie quickly found a single and gave us another perfect opportunity which we blew with the same skill level.

She looked over her shoulder, communicated something that wasn't very nice and went to the next single. She pointed. Steve and I started to walk up for the flush when Okie quietly reached down and picked up the very live wild (not released) bird and brought it to me completely un-harmed. She handed it to me and went back to get the next. The next day, I was out shooting at the skeet range.

More

The dogs involved were Okie, Sooner, Adam, Colleen (Golden Retriev-er), Robbie (Golden Retriever), Aussie (Weimeraner), Hershey (Chocolate Lab), Annie (Yellow Lab) and several small poo-poo dogs. I have stories I could share on every one of them, but by now, you get the idea. Keep your sense of humor and love them.

Colleen

The Dodge Power Wagon
Travis and Drew's First Vehicle

Elevated above normal with huge snow tires, she was an imposing vehicle. Three-quarter-ton, extended cab, with a full-size bed made her longer than most any other vehicle in the school parking lot, and her turning radius was intimidating, but she was theirs. She had obviously seen better times with numerous dents, rust, and bent bumpers, but she was beautiful to me even if the boys didn't agree. This was their first vehicle and she was virtually indestructible.

I first encountered her in a used car lot in Bozeman, MT, where we lived at the time. She had a large slide-in bed camper, but she looked like she could go anywhere. So, I bought her, sold the camper, and we had a hunting/fishing vehicle. To my knowledge, she never once got stuck in snow drifts (a small car would almost disappear in the snow before she started pushing snow with her bumper). Thus, other vehicles paved the way for us.

Once on an elk hunt, a young man was stuck in a snow bank so I pulled over to "help." I told him "I will pull you out backward but don't give it much gas or you will run into me." I started pulling and immediately he came free; showing typical young man intelligence, he stomped it

when he felt the tires grab. In my rear view mirror, I saw this car coming at me much like a torpedo with dead aim for my rear bumper. He hit me at approximately 20 mph which did dent my rear bumper maybe an inch or two (hard to tell, for there were so many dents already). Being quite a bit lower than mine, his car rammed under my rear bumper with devastating results. He was quite gracious, admitting it was his fault, but I sure hope he had a good insurance policy. The Dodge stood there looking down on his car much like a linebacker that had just decked a quarterback.

We moved back to Oklahoma a year later in July. The Power Wagon had no air conditioning, it was August, and her heater ran all the time (never seemed to be a big enough problem to have it fixed). It was 100°F when we crossed the Kansas border into Oklahoma. We never did get that heater fixed, and we had her for five years in the Oklahoma heat. On a hot day, we rolled down all the windows, turned up the country music, and went fishing.

The floor board was gone! Driving down the road, we could watch the road and gravel go by underneath our feet. I repaired that floor board one time, but it rusted out again in two years. I gave up, so as long as we owned the vehicle, we saw the road go by under our feet. A thick floor mat helped a lot, and the frame was very solid so it wasn't so obvious.

The spare tire was mounted to the front bumper much like a battering ram. In fact, we did frequently push with that tire. It was securely fastened, but rust has a way of loosening bolts. Once coming back from a deer hunt, we hit a bad bump and the tire took off. First, it bounced high from the bump, luckily I hit the brakes and watched the tire head down the road. Thank God no vehicles were coming for the tire and mounting frame bounced down the road for ¼ mile before going into the bar ditch. I fixed it by throwing the tire and frame into the bed of the truck where it stayed for the rest of the time.

It was completely impossible to drink a cup of coffee in the Power Wagon for her hard suspension would throw at least half of the cup back in your face. I never did learn my lesson, and continued to try, issuing a stream of bad language at every bump.

On one deer hunt, I guess I had forgotten to grease the wheel bearings on the front driver tire. I heard the noise of bearings rattling but I was only two miles from deer camp. Surely, I could drive slowly and make it to camp where I could remove the wheel and take it to town. Surely not! About ¼ mile from where I would have stopped, the wheel came off, and the truck settled down on the wheel, spraying sparks, luckily it did not start a fire, and my hunting buddy had a wrecker. We hauled her to town, replaced the tire and bearings (the wheel was all right believe it or not) and went hunting.

I believed in filing stuff carefully in the truck, so when we opened the door, there'd be old tire irons, five or six coffee cups, a McDonalds wrapper or two, coke cans, and just about everything else that might fit in a truck. Listen to Jerry Jeff Walker's "Pickup Truck Song." That was my vehicle.

Travis and Drew were good drivers and never once caused a problem, but I thought when time came for their first vehicle, they should drive the Power Wagon. They didn't think much of that decision but it was better than walking the 3 blocks to the school. Travis drove it one year; then he got his own pickup, and Drew who is one year younger took over. They found they became quite popular in that truck. First, if any school kid got stuck, guess who got the call? Remember, I already said I don't think the Power Wagon ever got stuck.

In the school parking lot, they always had a parking spot. In fact, they often had their own plus one on each side. The Power Wagon was intimidating and indestructible. All the kids knew that in an argument, their vehicle would lose.

Drew got his own vehicle one year later and the Power Wagon was discarded like she wasn't part of the family. I have always regretted that decision, but you know, she may still be running, some 20 years later.

The Motel Angler
Lessons Learned

On our way home from a great trout fishing trip in Montana, my wife and I stopped in Cheyenne, Wyoming. The boys had stayed behind in Stillwater nursing growing hormones. We rented a motel room and went next door to a restaurant for dinner. While there, we met "the Motel Angler." The lessons he taught me then are still with me 30 years later, and I want to share them with you.

An older person, he was significantly slowed by a stroke. His voice was strong but somewhat strained as he fought to put the words together. His hands and arms were still firm and muscular, but the trembling and lack of control showed the stranglehold a stroke can place on what was once a powerful strapping young man. Immediately, I liked this man, and I felt for him.

He overheard Kathy and me reliving our trip, and joined in the conversation. His wife, an obviously loving supporting person, watched and listened, mostly in silence. We wish now we had gotten to know her better, but he was fascinating and we listened intently.

He had been an oilfield worker in Alaska for many years in Prudhoe Bay and on the pipeline. He had been rough and tough and fished all over Alaska for grayling and salmon. Like most roustabouts, he had been a powerful, partying, hard-working man who had landed many large fish.

Then, the stroke knocked him down a bit. He retired, moved his wife and kids to Oregon and lived near a famous salmon and sturgeon stream. Exactly where is simply not important as you will see. Close to his house was a fast-water chute that funneled all the larger fish into a narrow area. Here large sturgeon and salmon would fight their way up- and perhaps downstream. These denizens of the deep didn't know it, but they were in trouble when this guy moved into town.

He couldn't cast his fly rod anymore, but he could back troll or rather "hang onto" a multi-trebled, hooked monster of a lure tied to a heavy line leading back to his bait-casting rod and reel that he would hang onto for most of the day. The outfit probably could have hauled in my pickup with the camper on it, from the way he described it.

He would cast that heavy-duty tackle into the water and back troll it adroitly into the chute. He knew that all large fish passing through the chute must get by this lure. When the fish encountered the lure, they would either smack it out of the way or savagely attack it. His hands and arms would recover their youthful vigor and strength as he managed a savage strike with an upheaval of the rod. The battle was on.

The rod and line did most of the work but he would hang on and slowly work the aquatic giant toward shore. If needed, and often it was, someone was always around to help him land this fish, for he was apparently somewhat of a legend in the area. One special fish, a sturgeon, filled his pickup bed. Others were simply huge. I imagine he's now fishing in Heaven and has regained all his youth and strength. I hope so, for I admired him immensely.

Most of us would have quit. He simply did not, and found a way to still "do it." He taught me a very valuable lesson that you probably see already.

I am now 73. Thank God no stroke has taken hold yet, but arthritis, aging knees, and other *minor* ailments have me working harder to keep doing it. I shoot a recurve with my right hand. The knuckles on my right hand swell significantly and hurt most of the time. Every time I want to complain, I think of this guy and what he must have gone through. A couple of IB and a warm pair of gloves are enough to keep me going. Kathy, our son Drew, and I still elk hunt out of our own camp in Colorado each year. It seems the elk move higher and the mountains are a little taller each year, but I now put on a knee brace and keep climbing. Drew gets to the top well before me, but I GET THERE. I will never quit, I will always find a way (like the Motel Angler did), and I will be grateful for the chance. I pledge to all of you that I will never quit, and I hope you join me in that promise.

We never saw that man again, and I have no idea what his name was, but he affected me, and I'll bet others around him, a great deal. I cannot thank him now; but just perhaps one of his kids or other friends will see this and smile. If you do, I'd love to hear from you. KEEP GOING.

> "The road goes on forever; the party never ends."
> —*Robert Earl Keen.*

> "If I live the life I'm given, I won't be scared to die."
> —*Avett Brothers (but I have some things to do first).*

Good-bye, Okie

WRITTEN ON A VERY SAD DAY WHEN OKIE, A VERY GOOD BIRD DOG AND MY BEST BUDDY, LEFT US; IT NEEDS EDITING BUT I CHOSE TO LEAVE IT AS I WROTE IT.

Sixteen years ago, Kathy and I carried you into our home—a warm fuzzy full-of-life puppy that immediately fell in love with us and we you. You never once cried even that first night. We all knew you were home. Tonight, Kathy and I freed you from your pain and suffering. You are now in Heaven, probably already pointing coveys. Your body, we carried to a hole in the ground and buried beside a pine tree that like the Phoenix will rise and ALWAYS remind me of you. I love you, Okie.

I came back inside, fixed a bourbon and branch, and now am therapeutically putting down my thoughts. When you died in my arms, your life rushed through my memories and tears. I remember the blue grouse hunt in Montana when you and I climbed to about 10,000 feet and finally shot some birds. You ran to retrieve, but it was 200 feet downhill. You got the bird, gave me a look that said something about my ancestry, and dropped the bird. After a short discussion, you decided to go back and get it. We proceeded to hunt some more.

I remember all our hunting, fishing, and just plain camping trips. You always liked to lie against me and were always on guard. Many times you growled long before a car or another person showed up, so I learned to trust your warnings. You were always right.

I remember your jealously and difficulty in honoring another dog's point; yet, you never once got into a dogfight. You felt you were the best, and by damn, I agree.

I remember your joy in the boys when they were young. You always wanted to play with and protect them, as they were your family. When they got older, you taught them how to bird hunt. Thanks.

I remember the way you would lie beside my boat or truck and wait. You knew I couldn't leave without your knowing it and, by gosh, I should always take you. I have a picture of your laying beside my Jon boat, waiting. I will cherish it always.

I remember your keying into the sound of my ole Remington 1148. Whenever I was going hunting, I would always check the action before leaving. You always heard that and would howl for hours if I didn't take you. I would pay $1000 to hear that howl again.

I remember you and me lying in front of the fireplace. You loved to fall asleep in my arms while I watched TV. Your warmth, breathing sounds, and presence always gave me comfort. Thanks.

I remember proudly the days afield when pointers and other "hell bent for leather breeds" would run circles around you to start the day. You were always still going at the end of the day when they were lost, dragging tail, or both. No dog could ever match you in finding singles, and that made me very proud.

I remember one day when you were young and wouldn't hunt. I kicked you gently to spur you. My heart ached then and still does. I don't think I ever struck you after that; but if I did, I was wrong. You gave me everything you had to give. No one can ask for more than that.

I remember your getting older and coming closer to foot. Fortunately, I was getting older also, for a couple of years toward the end you and I would sit on a rock, take a break, talk a bit, and then hunt some more. I believe that these might be my fondest memories.

I remember Kathy, the boys, and I watching TV or just plain loafing. You would always circulate from one to the other asking only for a little rub on the ears—life's greatest pleasure.

I miss you now and will miss you forever. Aussie is four months old and will likely join me in my trips in the future. She will never take your place, nor will the next one take hers.

I pray that you have 10-covey days for all eternity. I pray that the sand burrs will somehow miss your feet and that God will put you with Dad. You two carry a large part of my heart. Hunt together, and find a good lease for when I join you.

I don't want to stop writing for somehow, that's the end, but it isn't. You will be with me forever in the same way that Dad and I hunt together today. Okie and Dad, I love you.

The Important Things

At home, in front of the computer, for hours I struggle with the measurement of energy savings and meeting protocols.

Movement outside the window catches my eyes so I look. On the woodpile stand three bobcats working as a team.

One quietly crouched at the south end of the pile while the other two devilishly tear my carefully stacked wood apart on the north end.

They scratch and claw the logs, throwing them carelessly and loudly aside, making me wonder why.

Wham, the quiet one jumps and ferociously tears into a fleeing cotton tail, leaving a dead, bloody rabbit in the quiet one's jaws.

All three bound up the hill to a secluded spot to enjoy their hard-earned lunch, leaving me to wonder, why do I worry about the insignificant, when the important is just outside my window?

It's okay to embrace life and death.

Turkey School

Drew and I learned a few lessons during the last two seasons that we'd like to share. A couple of hunts, especially, show just how wary turkeys can be and how turkey hunts can be very successful even without a bird over the shoulder. Persistence does normally win the long run, however, as you will see.

I've been chasing turkeys, deer, etc. for many years and raised my sons doing the same. Drew is my younger son who lives with his wife and son in Pueblo, Colorado. We meet annually in Darouzett, Texas, to chase long-beards in the spring and again in September to chase elk in Colorado.

Darouzett is the home of the "Last Buffalo," an old bank building that has been remodeled into a hunting lodge. The lodge and the hunts are offered by Mr. Wade Robertson, his father, and sons. This annual event for the Turners has become quite an affair, as some years they are joined by their wives and Drew's son and friends. I'm going to tell the stories of two of these hunts. On the first hunt described below, there were more Turners in Darouzett than ever before in recorded history.

It was a windy moonless night as we tried to find our way (without flashlights) along a field beside the Beaver River. We knew there was a makeshift blind up ahead somewhere, and we stumbled over it in the dark. Two hen decoys and one jake made up the spread west of the blind as Drew and I settled down against two trees. He was shooting an old 1148 that my dad gave me, and I gave him some 40 years later. I was shooting a new black powder 12 ga. shotgun. The smoke pole was charged with 110 gr. black powder substitute, two wads, a wad cup, 110 gr. equivalent of #5 shot, all topped off with one more wad. This takes awhile to load, but the pattern is great and deadly out to about 35 yards.

At first light, the gobbles started. There were about 3 gobblers announcing their presence and some faint hen talk waffling through the cottonwoods. I gave faint tree talk back and they answered immediately. Through my face mask, I was smiling. Experiences like this are why we go to all the trouble.

After about 30 minutes of this communication, they flew down and started to strut, still out of sight. The gobbles went back and forth across the field, and we could picture that they were fanned and hot. Unfortunately, they also had hens, and no matter how seductively we might have sounded, hens in sight were better. We talked back and forth about an hour, and they went silent. I could picture that the hens were starting to go to nest. Drew and I also went silent, but we clucked and softly yelped every 15 minutes or so.

At almost exactly 10 o'clock (note how close this is to the old lore about the hens leaving them about 10), we saw a head stick up over a rise and look, with his neck stretched to the limit. It was a gobbler. We got ready, and he started moving our way (about 100 yards out). As he got closer, we could see it was a jake, but we kept up the soft flock talk. Every time I yelped or clucked, he fluttered, but didn't fan, and he moved obliquely across the field, never coming closer than about 50 yards. I was thinking that my calling must have the sex appeal of a slug. (*There is another explanation, as you will see shortly.*)

He ducked under a fence and headed toward the river, but before he was out of sight, one lone hen started toward us from about 50 yards south of where he first showed. She answered my soft flock talk with her own soft clucking and closed in until she was about 40 yards out. She dropped behind a sand hill and disappeared. A short time later, sand started flying in the air with a glimpse of a wing periodically. She was dusting herself and putting on quite a show. We grinned and kept talking; she was a great decoy.

Just out of curiosity, I tried a fairly demanding series of yelps. The hen stuck her head up, looked at the decoys, yelped back, and came on in. She stopped about 15 yards out and started scratching and feeding. I was silent, as she was simply too close to chance a call. After about 10 minutes, she stopped, stuck her head up, looked over her shoulder and issued a series of loud evenly spaced clucks. There must have been 10 to 15 in the series. It sounded like an old hen using an assembly call, but they were clucks, not yelps. Drew and I turned slowly in the direction she was looking, and **there he was**, more than 3 hours after we had started calling. He walked straight in with no hesitation, came straight to the decoys, ignored the jake, and started to strut. He was about 20 yards out.

Pulling the old "you shoot," "no, you shoot" trick, we finally alerted the bird, and he ran off. Pulling our thumbs out of our anatomy, we laughed at this silly mistake and really appreciated the experience. We gathered the decoys and headed to meet Wade at the truck.

Looking back at that experience with Drew and me, is it possible that the jake was afraid to come in because he knew "monster bird" was right behind? When he fluttered, he wanted to strut and gobble but after a beating or two; he knew better and ducked out of sight. Perhaps he was the first scout. Also, the hen was the next scout sent in to check them. She knew "monster bird" was right behind and she was protecting him. (Turned out he didn't need protection from these two Keystone cop turkey hunters.) I've not seen birds act like this before, but this is a true story. Interesting.

Getting back to my story, there is more, and this time a truly nice bird was carried back over a shoulder.

At about 11 a.m., it was getting very warm, especially for two turkey hunters with camouflage outfits. We joined Wade and suggested that we go to the "mott." West Oklahoma and Texas is mainly short-grass prairie with clumps of trees periodically. We call those "motts," and I have no idea whether that's a word or not. It is to us. Wade dropped us off about ½ mile from the mott. The walk through the pasture was very hot, and we were sweating profusely when we reached the trees. The large cottonwoods provided shade and cooling breezes that felt like an old friend welcoming us back. We liked this place.

We had hunted this mott three years in a row, and every year we had taken a bird or two. The previous year, I harvested a nice 3-year-old bird with significant beard and spurs. While picking up the decoys, I heard a loud gobble announce a newcomer. I sat down and started calling in the open (no decoys in place, no hide, and a dead bird at my feet). He marched right in like he owned the place (turned out he did) and went into strut about 10 yards from me. I would not shoot another, so I just enjoyed the show. He spotted an ugly out-of-place pile of leaves (me), broke strut, gave a disgusted cluck and marched out of there. That was quite a sight and (I think) the beginning of a long relationship. That's when the lore of the "mott bird" started.

Drew drove in that night, and guess where we were at mid-morning the next day? Sure enough, a gobbler answered our call fairly quickly; but this time, he was much more wary. He came in and stopped about 60 yards out. I tried to call him closer, but he tired of the game, marched into the prairie and gobbled. This became a pattern that we were to experience again, and Wade was to experience with a client after we left.

When the bird was 3 or 4 years old, I went to the mott one afternoon about 3 p.m. and tried calling. A bird answered from the sage brush, but he gobbled seldom and never in the same place. He circled about 180 degrees, gobbled two or three times and disappeared. About 7 p.m., I picked up the decoys and left, wondering if that was the same bird. Wade met me, and we joined Drew at the Last Buffalo. Early in the morning we went to the Beaver River, but getting back to the mott:

WE SET UP THE DECOYS and backed up against two large cottonwoods. We ate our oranges, napped a little, and talked a lot. We called every 15 minutes or so, and we tried to sound like a bunch of turkeys dusting in the midday. I sprinkled in some arguments just to make it interesting. *He gobbled!*

It was so quick; Drew and I looked in somewhat different directions. He was out there in the sage brush, but we weren't sure exactly where. We continued to call, napped a bit more, and chased shade around the trees for another hour or so. *He gobbled again.* He sounded off once each time and that was it, but this time we thought we had him located.

I called another hour or so (do the math, that's about 3 hours total). Drew thought he heard him again farther off, so we relaxed a bit. I moved to a much bigger shade tree and took a significant nap. I woke, ate an apple, and had just picked up the slate and diaphragm to call when Drew shot. This was startling and unexpected, so I jumped a bit. I looked over at Drew and saw one mature bird on the ground with three jakes beating the crap out of the dead bird as they often do. When I moved, they saw me and ran off.

Drew told me he was actually almost lying down with the gun pointed toward the decoys and his eyes closed. He opened them, saw the mature bird march straight into the decoys with no hesitation, shot, and the bird folded. Three times over two years, we had taken the bird that we had hunted, and we both felt excitement and remorse. He was quite a challenge.

The bird's beard was over 10 inches, and the spurs were over 1 inch. We estimated his weight, field dressed, at 22 lbs. They're all nice birds, but that one was special. We will never forget that bird and the thrill of hunting him so long.

> Just happened to Drew and the old man? Could it be that bird somehow learned that ugly pile of leaves could hurt? Thus, for the next four encounters over 2 years (remember Wade also experienced him), he hung up out of range and learned that if he stayed, the hens would eventually come to him. He simply had to announce his presence periodically. Furthermore, maybe he sat out there when Drew and the old man were calling, extended his neck and surveyed the mott with his powerful eyesight. He spotted the decoys and tried to get them to come out. Persistence paid off and he couldn't take it anymore. Gathering his jakes, he sent them in, but when they got close, he took over, coming in to end the relationship. What a hoot Drew and the old man had!

Old Homesites and the Stories They Tell

In my travels across Oklahoma as a bow hunter, I frequently find old homesites. Usually, I spend some time at each trying to imagine life back then and picturing the family that lived there. Why did they come, and perhaps more poignant, why did they leave? Hopes, dreams, and lost dreams all cross my mind. This little write-up conveys my thoughts at one such site east of Stillwater, OK, my chosen home. By the way, the deer did come—not close enough for my recurve—but I did watch.

The land rush to Oklahoma was a tsunami
        of people full of hopes and dreams.
With sweat, blood, and tears, they built homes
        where many dreams died or so it seems.

The old homesteads partially tell their stories
        if we but sit, listen and look.
A flower bed lined with red stones and a tired, bedraggled woman
        wanting flowers, perhaps to sit in and read a book.

Wood has rotted and stones have crumbled;
        yet, the outline of their home remains faintly.
A crude fireplace at one end warmed many cold nights and
        two small rooms at the other speak of children plainly.

Imagine over there, in the dirt close to the front door for safety,
        a small girl plays with her crude doll listening to her Mom
        singing a song so gay.
"You are my sunshine my only sunshine; you make me happy
        when skies are gray. You'll never know, Dear, how much
        I love you; please don't take my sunshine away."

The boy's torn clothes tell of his personality,
        a challenge to Mom and Dad.

An old 22, a broad grin on his face,
*   two squirrels and a rabbit tell of the day he had*

Dad, doing his best, and an ole mule
*   working hard to plow red clay.*
Some crops are made when hail
*   and critters don't take them away.*

Back to today, bits of china, vivid colors, broken, and still on the floor
*   are a sign*
that, she wanted some class in their lives but when they left,
*   not all was fine.*

Perhaps moving on to better places, but likely,
*   disease, accident, or the bank broke the family hopes.*
*   Broken, or realized, we'll never know,*
*   but for their sake, we can a bright outcome scope.*

Back to my hunting, I grab my bow
*   and check out the old fruit tree.*
Somehow, it still bears hard pears on limbs struggling to live,
*   attracting deer to a lunch seemingly free.*

Yes, lower limbs are empty
*   but higher up, fruit remains.*
Deer tracks and scat below imply
*   the pears will soon fall to the waiting deer on the plains.*

That elm, my ladder stand will hold;
*   Perhaps, I'll meet this family as my hunt unfolds.*

Wayne C. Turner
Stillwater, Ok

Very Close Encounters
with a Bull Elk

The weekend started off rather miserably in the Cookson Hills of eastern Oklahoma. Early fall and one of the first cold fronts of the year were bringing winds out of the northwest with a ferocity that only the plains can produce. This was a deer hunt that we had planned for about six months, and the cold front might stop the deer, but not us.

Clint Christenson and I, both of Stillwater, OK, decided to go on with our scheduled hunt since it was a special draw, and we had both drawn tags. We expected beautiful country, good camaraderie, and perhaps a chance at a deer. This was an archery hunt.

The Cookson Hills are a beautiful part of Oklahoma. Mostly hardwoods with some pines tossed in for color, the Hills are one of my favorite places to hunt. The area has been good to me, always producing a good hunt and often a deer for the freezer. The beauty of this hunt's surroundings was no exception. It was early fall, but the colors were already starting to turn, and the streams were crystal clear. I often found myself wanting to just look.

Early the first day, we went to the check station, got our tags, and drove to our prescouted site. It was a ridge that always has rubs and scrapes during gun season and that, until this day, had always given me deer to see. With the wind strongly blowing out of the north, we spent a great morning in two trees overlooking a nice trail. As the sky lighted in the east and day crept up on night, I snuggled deep inside my coveralls and stocking cap. My nose stung cold with each breath, but the discomfort was well worth it. The reds, yellows and rusts framed by the clear blue sky that normally follows a cold front were spectacular. Blue jays were screaming as they worked acorns; but there were no deer. We saw many squirrels, some raccoons, and lots of bird life, but nothing brown on four feet with antlers. We decided to scout some more.

Clint found a remote field with 10 to 15 very nice rubs in an isolated corner. That became home for Clint for the next 24 hours (with only a short night's sleep). I found another, less remote, field that had an isolated corner with lots of sign, some rubs, and a few early-season scrapes. I placed a tree stand about 10 ft up a pecan tree and declared it home. We spent the early afternoon listening to football and napping under a gigantic red oak. It was a great afternoon even though our team lost.

I was settled in my tree stand by 4:30. Since the field was exposed, I put on a full suit of camouflage. I had a face net, camo gloves, and normal camo overclothing. I was literally a slightly fowl-smelling bump on the tree. (The campground was primitive, so showers were not an option, and the creek was awfully cold). I normally use very good personal hygiene when bow hunting, so this time I liberally applied baking soda to try to destroy the odor. As I soon learned, it was good, but not good enough.

Around 5:15, a bull elk started bugling behind me on a ridge about ½ mile away. (Yes, Oklahoma has some elk.) I had applied every year for more than 25 years for an Oklahoma elk tag and never drawn, so I could not hunt him. I settled back and enjoyed the sound. No sound will raise your adrenaline, bring the hairs up on your neck and draw the predator from within more than the bugle of a bull elk. This sounded like a fully mature bull, and he was hot. I believe he was being challenged, and I was hearing two bulls. If so, they sounded close in size and were very near each other.

While it was a little early for our whitetail rut, it was perfect for the elk. He bugled every 10 minutes or so for about an hour as he and (I assume) his cows moved down the ridge behind me. I didn't hear them for awhile, but later, he started again (very close to me) and moved on across the ridge to the next valley. I thought, "Now, the hunt is successful; this is the first bull elk I have heard bugle in Oklahoma." I settled into my stand for the rest of the evening.

About 10 minutes later, I heard what sounded like an army coming through the woods. Deer don't travel like that, so I knew it was turkeys, elk, or a clumsy hunter. I honestly thought it was a hunter, but I turned in that direction, brought my recurve to where I could draw quickly, and watched. Suddenly, I realized the tree branches were moving. Then, a mature 6 x 7 bull elk materialized about 50 yards away. He was a beauty, probably scoring around 330, which was enough to truly excite me.

Already facing him, I settled down to enjoy the show, wishing I could draw, but knowing that this sighting alone was enough. He trotted straight toward my tree and got within about 15 yards when he scented me. He froze and was obviously irritated but probably too hot to care. I studied him carefully. He did not smell very good to me, but I think a cow would have loved it. His hide was ragged, muddy, and sweaty. He showed no blood, but I decided he had been arguing with the other bull and had lost. Thus, he was irritated and less wary than normal. At that point I understood why the other one bugled so much, and I thought, "Wow, I would like to have seen him."

The bull had smelled me and for some reason decided to follow my track. I had never seen one do this before. He followed my trail straight to the tree where I had leaped to the first limb, and he literally followed me up the tree with his nose by going from limb to limb until *there we were nose-to-nose at less than 30 inches.* Keep in mind, he was pretty tall especially with his neck extended to full length. I watched his nostrils pulsate with breath as he tried to figure what that smelly lump was, or perhaps he already had and was trying to intimidate me. If I had been on the ground, I would have yelled, and we both would have bolted.

The tree was my security, so I just continued watching. He studied me for about 5 minutes before deciding I wasn't worth the trouble. Then he walked under my tree and stepped out, quartering away at less than 10 yards. GEEZE!! I brought the recurve up, mentally picked a spot, mentally drew and automatically released. Mentally, it was a fantastic shot entering behind one rib and quartering forward through both lungs. He didn't make it to the wood before collapsing.

Truthfully, I was afraid to mock draw. As you instinctive shooters know, once you draw and anchor, the release is almost automatic, so all of the above was in my mind only.

He fed off another 10 yards to where the breeze from me hit him perfectly, he slammed on brakes, whirled and looked directly at me. He knew exactly what I was and where I was. He telegraphed messages that can't be put in print and again decided I wasn't worth the trouble. He put his head down and fed off until he was up the ridge and out of sight. I thought, "Wow, it doesn't get any better than that," but almost immediately the show continued. I heard the deer coming, before the bull was out of sight.

Farther up the hill, in a bad wind direction, I heard the tick, tick of an approaching deer. Of all the sounds in the woods, nothing is more distinctive than the dainty walk of a deer that I describe as ticking. A doe

and a large fawn came into view. They were feeding my way and there was one lane where I could possibly shoot before they smelled me. To make a long story very short, they fed for about 15 minutes and stepped out 30 yards (later pacing showed it to be 35 yards) from my tree in the "smelly direction" (see bull story above). She smelled something she didn't like, turned facing me, stretched her neck in my direction and once again, the nostrils were flaring. She snorted a couple of times, stomped her front foot, and tried the fake feed routine. She would put her nose down and look like she was going to feed and then immediately throw her head up and stare at me. That has caught me enough times over the years that I knew it was coming. Thus, I didn't move, and that confused her greatly.

Because the wildlife department was trying to cull does, the fawn was large, and I won't shoot yearling or other small bucks, I had decided to take a doe if given the chance. However, my recurve and I are less than 100% at 30 yards, so I never brought the bow up, and once again just enjoyed the show. A third deer then joined them. She was larger and barren—probably an old herd or lead doe, for once she hit the "smelly spot," she bolted and ran without snorting or showing any other signs. The young doe decided to leave also and followed the old doe up the hill.

The fawn went along reluctantly.

By this time I was shaking, but for some reason I still didn't move. That was great, for I caught movement, peripherally, then moved only my eyes to almost directly under my tree. There came a whole family of coyotes. Two were larger animals, and four were almost fully grown pups. The pups were playing with each other much like children on a playground. They would run each other down and tackle; then the tackler would run the other way while the tackled one became "it" and chased the first. I must admit that I could have shot, but I was really enjoying the show. Thus, I watched some more. This went on for about 5 minutes as they played down the edge of the field and disappeared from view.

It was sunset. Never before had I been so near so much wildlife in a short, intense period. I reflected on this quite a bit the rest of that day and have since then. A kill would have stolen from those memories, and I will never forget that afternoon. However, don't get me wrong: if I had a bull tag, the wooden shaft would have flown, and hopefully, the tag punched.

Waiting until well after dark so as to not spook any deer that might be nearby, I climbed down my tree and hiked to the truck. Clint was waiting when I got there. I told him, "They write stories about the type of evening I just had." I guess I just did.

Yellowstone Fly Fishing Trip

7/28/98 Tues.

I left home at 6:30 a.m. and drove to the airport in Tulsa. A hot sultry morning greeted me as I loaded the truck. We're in a string of 100+F days. This trip to Wyoming and Montana in pursuit of trout will bring the added blessing of cooler weather.

A few months ago, my son Drew called and asked if I would join him and a friend, Seth, backpacking and fishing across Montana and Wyoming. A quick check of business schedules and consoling a very understanding wife who would love such a trip led to this.

Drew is an avid outdoorsman. He and I have fished and hunted across many states and have a relationship that only such strong shared interest can cultivate. We understand and love each other very much. The other day I thanked him for sharing his youth with me and sincerely meant it. He is getting married in September to a wonderful gal, Melissa. This trip is sort of a "last time under these circumstances."

They're driving from Pueblo, Colorado, and I'm flying from my house in Stillwater, OK, to our joining of paths in Jackson, Wyoming. Drew used to work at a fly shop in Jackson, so this will be a homecoming of sorts for him. I'm anxious to meet Seth who is also an avid outdoorsman and one of Drew's best friends.

7/29/98 Wednesday

Drew and Seth met me yesterday about 4. We drove northwest of town to a spring creek. I worked several rising cutthroats and had one hit. Gorgeous country with incredible views but fish with excellent eyesight. I later figured out that they were spotting my tippet. Gotta work on that float some.

Today, we got up early and ate breakfast at Bubbas. Bubbas has great bar-be-cue and an excellent breakfast, so we pigged out. We drove into Yellowstone, got our permits, and started looking for a campsite. I recommend that a person get an assigned site before going into the park. We looked and looked; finally, we found a site at Tower Junction. Then, we went below the Gibbon Falls and fished the Gibbon River. We fished up to the falls. They each caught a nice trout and several small ones. I caught two or three small ones.

Then, we drove back to the campground and fished the Gibbon River in a bog right below the camp. We each caught one or two small fish (all cutthroats) and sank up to our keezers in bog. We ate dried chicken fajitas (actually, much better tasting than the name implies) and went to bed. We left the cooler outside and then remembered you shouldn't do that in Yellowstone (bears). We were too tired and not worried enough to move the food, so we left it. Luckily, no wandering bear found us.

7/30/98 Thursday

We got up early and drove to Old Faithful. We took pictures of a young bull moose on the way. Seth saw OF go off, and we drove on to West Yellowstone and down the mountain to the Madison River Valley. This is a beautiful river valley with very dry countryside going right up to a world class trout stream. Unfortunately, whirling disease has hit the Madison hard. I fished very hard and had two small rises right up against the shoreline, but no takers. They caught two or three small ones. We quit about 9 and drove to McAlester where we ate at a very Montana-type steak house. The food was great and the atmosphere better. We drove on to Manhattan where I had a cabin rented at Twin Rivers Ranch.

7/31/98 Friday

I got up early and made coffee. The cabin is right on the East Gallatin River, another renowned trout stream that Drew and I have frequently fished. As I made the coffee at first light, a flock of geese landed on a gravel bar about 20 feet from the cabin door and proceeded to talk as geese normally do. When I opened the door, they quickly took to flight with quite a noise. I took a walk with a cup of coffee and saw sandhill cranes, muskrats, geese, several kinds of ducks, mule deer and whitetail deer. Twin Rivers is quite a place, and Montana is what God sent to me to let me know what heaven is going to be like. I really enjoyed this morning.

Today, I met with Mary Ellen of Gallatin River Ranch and made some decisions on lots (we might buy a place there). I went back to the cabin about 1:00 p.m., had a sandwich and took a brief nap. In the meantime, Drew and Seth fished Willow Creek and the Madison again. Any of you knowledgeable in world class trout streams will recognize all these names.

8/1/98 Sat.

We got up early again today and drove to Bozeman. We stopped at River's Edge and visited some. Of course, we bought some more flys. Then we drove to Livingston and started up the Paradise Valley toward Yellowstone's NE gate. We tried to get a site at Mammoth (just inside the gate), and they were wide open. Thus, we decided to take a chance and drive on in (mistake!!). We drove to Slough River Campground. There were no sites available, but there were some excited people. A grizzly bear had visited them early this morning. Now we have a dilemma. We decide to hike on up Slough Creek to one of the three meadows and fish. The hike is about an hour to each meadow. So, we got to meadow one and fell in love with it. There were fish rising in the stream and it was getting on fairly late by now. Thus, we decided to stop and fish.

We hiked about 2/3 the way up the first meadow and dropped down to the creek. As we were rigging our lines, we noticed a guy upriver catching fish. He caught three while I was watching, so I went by to talk. He gave me some great advice about the float and we talked awhile. He is from Virginia and knows Hank Norton well. Hank is a friend of Kathy's and mine and is an avid fly fisherman on the Chesapeake Bay. This guy makes his own bamboo fly rods and is quite an angler. I really liked him.

The valley was spectacular—clean water in the middle of a beautiful meadow against a backdrop of mountains all the way around the little meadow. We proceeded to catch lots of fish. Seth caught the biggest (around 18"). Drew caught one 18" and I caught several smaller. We heard what could have been wolves, and this is an area where the wolves concentrate. We liked that. We hiked out about dark and drank a beer. We agreed this was the best of the trip. We drove on to Jackson, getting there about midnight, and finally found a room about 2 a.m.

8/2/98

Up about 8 (short night), shower and back to Bubbas. Another great breakfast fueled a short trip over to Orvis where we spent some more money. They drove me to the Elk Refuge Inn and they took off back to

Colorado. All of a sudden, I was pretty lonely. I enjoyed being with them. In fact, I got so lonely I had to go fishing.

I rigged up and crossed the street to the Elk Refuge and Flat Creek. Flat Creek is a gorgeous, slowly meandering creek through a lush meadow with lots of terrestrials. The fish were rising, so I rigged a #16 drake with a small bead head dropper, and they TORE THEM UP. Fish were everywhere. They eventually totally demolished that fly. I would estimate I caught somewhere around 20 cutthroat trout in about 2.5 hours. WOW! I went to change the fly and thought I can't beat this so why not quit?

I did. While there and while walking back, I jumped quite a few woodcock. When I got back, I ordered a pizza. While waiting for the pizza, I finished writing this. Ate the pizza, drank a beer, called Kathy and told her I loved and missed her, and went to bed.

Tomorrow REALITY. I travel to Atlantic City, NJ, to teach a course. YEUCH!!!!!!

Two Duck Decoys

The old man passes two old duck decoys sitting on the shelf as he walks toward his word processor. Smiling, he picks one up and gently caresses the hull, much as one would pet a favored hunting dog. These decoys belonged to his two boys when they were younger, and they will belong to their children when the time is right. The memories they trigger are intense.

Once, many years ago, he came home from a day at work to find his two very young boys in the bath tub with RUBBER DUCKS. He immediately drove back into town, bought these two cork mallard decoys and brought them home. Thereafter, when the need for a duck in the bathtub arose, these decoys satisfied the need (at least when he was around).

The boys have now grown up. One lives in Stockholm, Sweden, and the other in Pueblo, Colorado. The old man misses them a great deal, but he is extremely proud of them and grateful for the experiences they shared.

The decoys bring back many fond memories as the old man holds and reflects on them. Perhaps the oldest memory is of Drew (the younger son) and his first duck. At a deer camp in southeastern Oklahoma, Drew was not yet allowed a rifle and sat in a tree stand next to his dad (the old man) for many hours, several harvests (over time), and many great memories. This one time, Drew tired of sitting in a tree stand and asked his dad if he could take his .410 and sit by a pond for a duck. Normally, his dad would not let him go after a duck with such a small shotgun; but Drew had been very patient and was an excellent camp partner. His dad said, "sure." They reviewed gun safety, chose 3-inch shells to maximize the opportunity, and went their separate ways. Of course, Dad worried the whole time for this was a first, but he did not show it (hopefully). Dad went to a close tree stand and sat to wait.

After about an hour, the .410 spoke. Its voice seemed to be about three times the normal volume, but the dad fought the urge to run and stayed in the stand. Then, about five minutes later, the gun spoke again. Dad smiled, figuring the son had a duck down and was finishing the job. Five minutes later, the gun spoke again. This repeated itself four times before silence ensued. Upon returning to camp, Dad found a very proud smiling young son there with a very wet looking common merganser. A merganser is a diving duck so every time Drew shot, the duck dove, explaining all the shots. Dad was as proud as the son was of course; but then, a problem arose.

Common mergansers are not the tastiest duck around; in fact, they taste like rotten fish. However, Dad worked hard to develop an appreciation for life and responsibilities in his sons. Thus, Dad and Drew cleaned the duck and carefully placed it on ice for the trip home.

Immediately upon arriving home, Dad dropped Drew off at the house and said he had to go to town. When he came back, the merganser had put on significant weight in the cooler and looked much like a domestic duck. Luckily, Drew did not observe this marvelous change and proudly showed his mom the duck. The whole family ate on that miracle duck that evening and all commented on Drew's hunting prowess. There was a mysterious small grave at the end of the driveway that no one else ever noticed and that caused Dad to smile every time he drove past for some time.

Right behind that memory is one of a hunt with the older son Travis. Before moving to Sweden, Travis had come home for a visit late in the duck season. He and Dad (the old man by then) had gone to a duck marsh they had built and managed for some years. The marsh was a favored gadwall hangout, so they always did well and got to visit. This time, even the gadwalls had "smarted up" and were quite decoy/call shy. A windless day with no decoy motion on the water made it even worse. Travis and the old man called, pleaded, and begged, but all this one flock would do is circle and look. After the fourth pass, the old man whispered, "They don't like not seeing motion." On the next pass, he threw a rock into the middle of the decoys. The response was immediate and comical. The whole flock saw the motion at once and pitched straight into the stool with no warning. The first feet were wet by the time Travis and the old man could react, but react they did. Once again, the whole family ate well that evening. They had this last hunt before he moved across the big duck water (Atlantic Ocean), and they talked a great deal. The old man could tell you most of that discussion even today.

Then, many memories flew by as the old man remained there thinking. Once, Drew, he and a friend were hidden on a pond dam working a flock of ducks into the stool. When the birds finally pitched, the old

man quietly said, "All right boys, they're mallards. Pick the green heads."
When the first feet hit the water, the guns spoke. After the smoke and
noise cleared, four green heads lay dead on the water. The Golden (who
in essence was a third child), was bringing them in one by one. It's easy
to make a mistake when a whole flock comes in like that, and often stray
shot brings a Suzy in by accident. This time, Providence smiled, and only
males were harvested. (Suzies were legal up to one per hunter per day).
For some reason, the old man remembers and savors the moments of the
flock working them and the time he spent sending the dog and watching
the retrieves. For reasons that only hunters can understand, he does not
recall the actual harvest.

Once, the boys were on the bank of the local river because all the
ponds were frozen and the few ducks that remained had gone to the open
water. They had harvested only two or three birds, but the banks of the
river and the surrounding hillsides were white with clean snow, and the
ice in the river moaned and cracked as it flowed downstream. The decoys
swam seductively in the current behind a log jam that blocked the ice
floes. The visual and sound effects of all this remain with the old man
today. He closes his eyes and could conjure up again that feeling; but he
shutters as he remembers the cold. It was right at zero, and they didn't stay
long. He got up and walked out to pull in the decoys. He had the first one
up when Drew whispered loudly, "Dad!" Knowing immediately what was
happening, the old man bent over and started calling. He could see the
flock's reflection in the water and they liked his call. Immediately, they
dipped for a better look. Using his normal tactic, the old man waited until
they were not directly overhead, and he called again. Because open water
was scarce and valuable, the ducks did not want to leave, but what was
that ugly blob in the middle of those birds? It looked like a gray-haired
old man, but surely not!!! After a few passes, the birds came in (away from
the old man) for one more look. Knowing they were now close enough
to shoot and that the ducks would see the old man on this pass, the boys
fired twice. The old man, still looking down at the water, saw the birds
coming toward him, saw two immediately fold and drop straight *at him*.
They landed close enough that he could simply take a step and pick up the
two widgeons that the boys had taken. That memory is vivid.

Not having enough time to relive all the memories the decoys brought back (and because, by now, you have the idea), the old man carefully placed the decoy back on the shelf. He wiped a tear of remorse, gratitude, and pure love off his cheek, sat down at the word processor and wrote a story he called "TWO DUCK DECOYS."

This story is dedicated to Travis and Drew who are now leading their own adult lives. I am very proud of you and THANKS FOR SHARING YOUR YOUTH WITH ME.

The Old Man

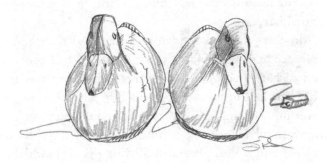

Memories of Hunting and Fishing with the Boys

"The road goes on forever, the party never ends."

Now it's my turn to raise Travis and Drew though I long for the peace of
    Dad's arms
Teaching them to love the wild as I try to protect them from harm
Hearing their first turkey gobble or elk bugle call
Hoping they get struck by it all
Frog gigging, trot lining, fishing, bird hunting, bow hunting all part of a
    plan
Helping them each turn into a sensitive yet strong man
Here are some of their stories
As we hunt hard for maturity and memory glory
These are memories of hunting and fishing with the boys

The boys and I frog gigging a neighbor's lake
As they learn this can be fun and fried legs and taters taste great
Sneaking over the same lake dam, wishing for luck
Green head mallard flushing, Drew shooting, down comes his duck
Getting home, Drew excitedly babbling his story
To an understanding Kathy who embellishes the glory
Travis getting his first large buck on Cockrell's ranch
Shooting an older-than-he model 94, 30-30, stand hung on a tree branch
These are memories of hunting and fishing with the boys

Drew, on a Colorado mountain stalking through dense timber
Bow in hand, trying hard to be stealthy and limber
Bull elk standing, exposing vitals, drawing and shooting
Arrow striking true, downing his first elk, Drew coming to camp "hooting"
Travis, Kathy, and I hiking to a high mountain lake
Finding many large cutthroat waiting for a fly to take
Travis, a mature adult by now, but still our lad

Saying "Dad, thanks, that's the best fishing I've ever had."
These are memories of hunting and fishing with the boys

Adopting the boys, our bird dog Okie somehow realizing her job was just
To protect the boys, teach them to hunt, and meet the school bus
Teach them she did, as today they are healthy adults and many a quail fell
I only wish I had done my job as well
As they got older, Kathy, the boys and I on Lake Powell chasing stripers,
     catching many, in our bass boat
Most importantly, their still wanting to be with us though I sometimes
     seemed as dumb as a goat (luckily, I grew smarter as they grew older)
These are memories of hunting and fishing with the boys

Why was our doing this in the great outdoors so important? For as Robert
     Service said:
"It's the great, big broad land 'way up yonder,
It's the forests where silence has lease;
It's the beauty that thrills me with wonder,
It's the stillness the fills me with peace."

Here's another by Steve Walburn (p7 *Grays Sporting Journal* July 2015)
Long ago I found something indescribable in those woods and I took
     away more than could ever be repaid.
Travis and Drew, thanks for sharing your youth with us.

Part III

The Old Man Grows Older
In the Great Outdoors

The boys moving on and living their own lives didn't start abruptly; it just phased in or happened in surges. About 2000, both had started their jobs and their lives apart from Kathy and me. They are still and will always be our sons or children, but they are really quite impressive adults which may have occurred at least partially due to the outdoors influences.

Then, however, I found myself hunting more and more with friends and sometimes alone. There were difficult times: One time in Pawnee County, Oklahoma, the bird dog and I were talking while driving to a hunt, and I looked at the empty front seat in the truck. That's where one of the boys had always sat. For some reason, perhaps because I loved and still love them, I started crying and pulled over to the side. I wrote them that night. I think that's the only time I actually cried, but gosh, they had been a very essential part of my life.

That experience is why I wrote this book in three parts of my development. The fourth is a hunt in Africa that Kathy and I recently took. It just seemed to call for its own part.

As I write this, it is late May 2015 on the Chesapeake Bay of Virginia. Kathy and I retired in 2008 and moved to Colorado half the year and Virginia the other half. We live the summer and fall in Colorado with enough winter to enjoy skiing. In early March, we move to Virginia where we are licensed crabbers, oyster farmers, fisher people, clammers, and kayakers on the Chesapeake Bay. What a life we have, but hey, I wish that Dad and the boys could enjoy it with us. Actually, both boys spend a couple of weeks each year here on the bay or in the mountains of Colorado.

This period has been extremely productive in my search for finding Wayne. You will see I talk less about the hunt and more about the spiritual side of the experience. I will always hunt and perhaps harvest some along the way, but there are fewer kills and more discoveries. You know, if I'm not careful, someday I will be educated and understand all this. I hope not, for the journey is too rich.

Please enjoy this part.

Wayne Turner
Grimstead, VA

Frog Gigging and Life

The old man drives his pickup down the country road. The gravel under the tires sounds like popcorn in the background of the country music on his radio. For the first time in many years, he's going frog gigging alone. The pickup's lights pierce the darkness, illuminating specific spots briefly, much as his thoughts pierce the darkness of memory and bring back previous frog gigging trips. He has done this many times before, but always with his father and brother dating back over 60 years, or his own boys who have been gone for ten years. They are busy making their own families and lives now. On other trips, he has taken his students (as their college professor) and many times just kids in the neighborhood. This time, he wants only the company of his thoughts.

He stops the truck at the pond bank and steps out. Immediately, his senses are assaulted, as he knew they would be. The Oklahoma heat and humidity of late August make the air feel thick and heavy, much as the steam in a steam room. His nose picks up the earthy smells of mud, muck, and water. They smell good and make him feel at home. Next, he notices the sounds—first, the cicadas, singing their typical August tune. They are loud and shrill, actually painful so that he almost pulls back into the truck; but then he hears the deep resonance of a bullfrog bragging about his prowess. That sound has often been written as ribbitt, but somehow that doesn't do it justice. The call starts deep and gets deeper as the bullfrog expands his throat and issues both a challenge and a love call to the other sex. The old man loves that sound and stops for a while to soak it in.

Off in the distance, coyotes are greeting the evening with excited yipping that sounds like kids laughing on a playground. The old man thinks that the trip would be worthwhile to just sit here and let his senses absorb all this; but no, he is here to catch some frogs.

He gathers his equipment, which nowadays is nothing more than a strong headlamp and a gunnysack. One does not frog gig to kill frogs; so, he has chosen to only use his hands tonight. In years past, he would also have a frog gig. Ortega Y Gasset said, "One does not hunt to kill; one kills in order to have hunted." Replace the word "hunt" with "frog gig" and you have the idea.

He chuckles as he remembers his dad and his brother leading the way many years ago. They all had flashlights, gunnysacks and gigs. Often, they would fill three gunny sacks half full of frogs in an evening and then go home to fry a batch late at night. The rest were frozen. To him, nothing ever tasted better than frog legs and fried potatoes late at night at home with his brother and dad. Dad is now gone, stopping the accumulation of memories, but not dampening the intensity of so many trips frog gigging, fishing, hunting, and trapping together. His dad has been gone for 25 years now, but he still misses him deeply. His brother is still alive, and they are great friends; but his brother "grew up" and stopped doing such childish things. The old man chuckles as he repeats his motto to himself, "Grow Older, Not Old, and Never Grow UP."

Then, his memory leaps ahead 30 years to his own boys. He started them hunting, fishing, and frog gigging when they could barely keep up with him. He stops and laughs when he remembers the time a local rancher saw the light and thought he was a rustler. He and three young boys (around 5 or 6 years old) were working a pond when the hillside came alive with car lights, flashing police lights, and sirens. The police cars raced down the hill to catch them rustlers with the flashlight. The cars slammed on brakes and the police jumped out to see, standing in the mud, a young father with a gunny sack, frog gig and three young and thoroughly frightened boys hiding behind him. The sheriff took one look and started laughing so hard that he couldn't talk. The rancher jumped out of his truck and started apologizing profusely. The old man and his sons still had their fog legs and fried taters that night, but they also had something to talk and laugh about many times afterwards.

He hears another bullfrog in a different part of the pond; so, there will be some action. He pulls the ball cap down tightly as it gives some protection from the ever-present mosquitoes, and feels the cap tighten about his sweat-matted gray hair. (He chuckles—at least there is some hair left). He steps into the water. The warm water wraps his jeans tightly against his legs and he feels his old tennis shoes sink deeply into the mud. He likes this feel and is certainly used to it. He walks slowly around the pond with the light seeking the welcome feedback of two eyes on top of a black heap. At the first one, he is excited and ripples the surface enough that the bullfrog explodes back into the water. With a gig, he could have had that one, but with his hands, he has to be very close.

A little later, he sees another. This time, he collects himself and slowly moves into position. His flashlight is in his left hand shinning directly into the frog's eyes, transfixing it in position. His right hand is carefully reaching behind the frog until he gets within 2 or 3 inches. He grabs the frog behind the front legs and hangs on tight as the frog doesn't especially like this situation and is trying to escape. The old man dispatches the frog as his dad taught him, drops it into the sack, grins like he just bought a T-bone steak, and starts for the next one.

There, on the shore is another set of eyes, but this time it is a coon exploring for crawfish and just about anything else he might find. The coon doesn't hold tight for the light, but he presents quite a picture as the light shines on him running up the bank. Before the night is over, the old man sees another coon and a beaver that can't figure out what would be on the shore at that time of night. The beaver slaps his tail four or five times before becoming bored and moving on.

The old man takes six frogs (enough for a meal for him and his wife) and quits. Bullfrogs are somewhat scarce now, so he limits his catch. He has probably stalked fifteen different frogs to get those six, but now he has dinner.

He walks back to the truck. The tennis shoes and jeans are filthy with mud and soaking wet, so he squishes as he walks. Somehow, even that

seems right. At the truck, he changes shoes and pants, turns the light off, and reaches into the glove box for a small bottle of good bourbon that he brings on these trips. Into a tin cup he pours a small amount that he waters down with his ever-present water bottle. He calls this "saluting the end of the trip." The old man takes that one drink, and thinks how life does not get any better than this. He thanks God for his dad who taught him all this, his brother and sons who have often been with him on these trips, and his wife who somehow understands he has to do this.

He chuckles as he remembers the first time his wife (girl friend then) went home with him. She was from Washington, DC, and was/is cultured. He often says he has just as much culture; it's just a different type. She went frog gigging with him and his dad. She sat in the truck and watched the lights move around the pond. When they got home, he cleaned the frogs, and his dad got two pans of grease hot. He knew what his dad was going to do, as he loved to get the grease hot and ask a "newcomer" to drop the frog legs into the grease. The legs commenced to jump when they hit the hot grease and his girl friend screamed and jumped halfway across the kitchen. They married shortly after that and have been married for 46 years. She still talks about the jumping frog legs and now will even pull that trick herself.

The drink finished, he cleans the frogs, ties the legs into a freezer bag and drops them on ice. At the richness of it all, something seems to jump inside him also. The two of them will enjoy their frog legs and fried taters tonight.

Thanks, Dad, Travis, Drew, Ken and Kathy.

The Mystic Side of Nature

Tom Brown calls it the "energy that moves through all things." All indigenous tribes believe that God (Wakan Tanka) lives in all trees, animals, plants and even rocks. John Muir believed that nature is the best cathedral and that each tree that is cut to build a church drags us farther from God (my words but his feelings). I'm not that smart, but I know I feel closer to God in a tree stand than I do in a church constructed with small windows that keep God out. In the stories below, I am going to share some mystic experiences I have had while hunting, fishing or just out in nature. Hope you enjoy.

The Doe that Sacrificed Herself for My Family

The Oklahoma skies had spilled rain for several days, and I had cabin fever. It was early October, bow season was in, and I had my recurve and arrows all ready to use. When the rain stopped in early afternoon, I took my recurve and feathered arrows to a tree stand about 1 mile from the house. The stand was in timber at the southeast end of a large wheat field that had been replanted and was starting to show green which the deer love. I climbed the tree, tied myself in and said a little prayer, thanking God for the opportunity.

The sun came out and the wind died completely. It was one of those spectacular October afternoons with temperature around 50°F, no wind, and clean smells. Then, a large fox squirrel peeked around the trunk of a big willow and looked straight at me. As you probably know, they love to sit there and bark at you for what seems like hours, ruining the moment if not the hunt. This one did something strange, he came to my tree using the overhead approach, climbed down to the limb holding my tree stand strap and *sat on the limb about 1 foot from my shoulder*. He looked me in the eyes and looked back at the deer trail which I was watching and then just sat there. He definitely was communicating with me in some fashion. He sat still for about 15 minutes (really, time became a non-dimension to me as I was astounded), looked at me again, jumped to another limb, and scurried off. I shall never forget that experience.

Very soon after that, a doe walked out on the deer trail and stopped broadside at less than 10 yards. I picked a spot, drew my 58 lb recurve and sent almost 700 grains of arrow and broad head. The hit was ideal, going in high mid-lung and coming out low mid-lung with a complete pass through. The doe never looked up or bolted. She took one step forward, turned a complete circle, and died. I have a little ritual that I go through thanking the animal, so I did that, field dressed her, went home and thought long and hard about that experience. I am still thinking, some 10 years later.

Many indigenous tribes believe (again my words expressing their belief) that "you hunt an animal until the animal finds you." I am absolutely convinced that hunt was ordained in some fashion and that the squirrel was doing his best to communicate with me. WOW.

Another Squirrel That Liked Me

Once while hiking a trail in Utah, a squirrel came to my feet. I froze and leaned against a rock because he was staring at me as if trying to communicate. He jumped up on the rock, walked to me, and sat on my arm. Now, this is not a city park, this is a hiking trail in a mountainous area, so I was quite surprised. He sat on my arm and started playing with my jacket. Tiring of that, he leaped down and walked off. (My wife said he was eating what I had dropped during lunch, but I didn't have my jacket on during lunch.) Was he picking up on some energy flow like the squirrel in the story above?

I have a friend that has often had a bird land on his shoulder. He is Indian; the bird is a yellow flicker, and his chosen bird as a child was a yellow flicker.

What caused those moments?

She's Drowning

She surfaced by swimming hard, was able to get a breath, and then she went down again. Every few seconds, she would repeat the effort, but it was obvious the muskrat was drowning as she was tangled in nylon fishing line carelessly and innocently tossed aside by some fisherperson.

We were at a small fishing dock located in our home town. The lake was stocked with trout and is free to visitors and locals, so it was a popular fishing spot. The line was tangled in the dock pilings, and the muskrat was

tangled in the line. I had to help her.

Grabbing a two-by-four lying on the shore, I was able to reach deep in the water and break some of the line. Importantly, as you will see, to do this I was within 6 inches of the struggling muskrat. Breaking that line enabled her to stay on top and breathe, but more line held the animal tight. By reaching deep into the water with the two-by-four, I was able to break the remaining line and the muskrat was freed.

Expecting her to swim away, I kneeled on the dock to watch. Instead of swimming away, the animal swam to the dock and crawled up on the dock between my legs so we were once again 6 inches apart but now voluntarily. The animal looked at me deep into my eyes and then started grooming herself.

Completely enthralled, I did not move and hardly breathed. As a former trapper, I had been much closer to muskrats, but under completely different circumstances. Time stood still as the animal continued to groom herself right beside me. Kathy took a couple pictures, but we were both completely captivated by this wild animal choosing to be so close to me.

After a while (I have no idea how long), she slowly dove in and swam off, leaving quite an impression on us. I have no doubt that she was thanking me in the only way she knew how. I will always defend a person's right to trap in a humane fashion, and as a young man I did. Probably, my personal trapping days are over after this experience. Once again, I have no doubt she was communicating with me.

Circle of Life (Epilog)

All indigenous peoples believe that life can be represented by a circle. Picture, if you will, a circle drawn in the sand. On that circle are the deer, muskrat, birds, and squirrels talked about above as well as all life, including the ones we don't particularly want around (termites, spiders, poisonous snakes, etc.). I personally believe that plant life is also on that circle. Being on a circle means that all life is equally important. My life is no more important than the muskrat and deer above or even than the termites, spiders, and poisonous snakes. Very importantly, some life (such as you and I) must take other lives in order to survive. That's okay, as long as we show proper respect and honor the lives that we take.

If a poisonous snake is in the yard around my grandchild, I will kill it and then apologize. To honor it, I will likely tan and keep the skin. If that same poisonous snake is in the woods and is not a threat to those that I love, I will likely take a picture of it and leave it alone. Termites and spiders are more difficult to explain, but there is large spider (zipper) that builds a large web complete with a zipper-looking weave in the middle. I never disturb these very large spiders, even around my home, for they actually kill and eat other spiders and never seem to bother us. Frankly, they do bother our guests, but they can handle it.

When I'm mushroom hunting, I will take a few moments and pray to Wakan Tanka, who happens to be God for me, to let me find and honor those mushrooms. I try not to pick the first one, but I will thank it and I try to leave some for future harvests. If you don't do this, I believe they will hide from you. Of course, you could explain this by saying I have slowed down and become observant, but I like my explanation better.

This is what the circle of life means, and it took me 70 years to get to this point. I'm still learning, thank goodness.

Stripers in the Lights

The normal afternoon thunderheads were building as we watched them carefully, but they just evaporated about 1 hour before dark. Launching, we were soon a mile north of our home (close to shore) in our Hobie pedal fishing kayaks, catching large croakers on jig and soft tails. I filleted the large ones and put the small ones in our crab pots so they were welcome. As we fish through grass beds on the way to the point, we often catch speckled trout, sometimes large ones, but that night they didn't show. We were headed toward a local point that separates the Chesapeake Bay from our local small bay, creating a bar over which bait fish are swept during tidal flows.

We like to arrive at the point about dark, for there are two homes that have dusk-to-dawn lights on their dock, and the stripers love to feed under those lights. We were stalking and hoped that we could get into some larger ones. I switched over to heavier tackle (12-lb test) tied to a large minnow type lure, but Kathy stayed with her lighter weight (8-lb test) tied to a 3/8-oz jig with a soft tail. I was hoping for large stripers. She was hoping for fun.

The Bay was at her best. With only a light breeze from the southwest and on the lee side of the point, the waters were placid. The western sky was still glowing a dull red and the stars were out. It was almost worth it to just watch, but no, we were here for stripers and blues. As dark overcame day, the waters started with light splashes, and I began to catch schooling blues (1 lb or so) but occasionally we heard keerspalsh as a larger predator slashed through the bait fish. We were watching the lights all that time sitting off about 100 yards as we didn't know their approach path. There, 4-inch bait fish were jumping out of the water under the lights as the show started.

We didn't see the stripers, but we certainly saw the results as bait fish boiled trying to escape and a large wake was cruising through the lights. Time to go to "work," we pedaled (these new Hobies are awesome with pedals so that both hands are free) to within casting range. Kathy, with my admiration, was still using the lighter weight with jig and tail. I had my elephant gun (heavy tackle). On the first cast into toward the dock, I got a take as my lure passed under the lights. The take was surprisingly soft, and I thought I was back into schooling blues, but once the fish understands he is HOOKED, he explodes with a run that strips off about 20 yards (remember, I'm not light lining). After a short fight my lip gaff held him secure. (Please be careful, stripers have very powerful necks and will throw the lure hooks into your hands if you are careless). He was a schoolie striper that thought he was bigger. About 20 inches, he must have weighed 3 pounds. I released him and waited for the next show of life. Kathy and I both continued to catch fish for a couple hours. We let the fish grow up as we continued to wait for the big predators to show. Most casts yielded a strike and many a fish. The biggest was about 22 inches, so we never found the hogs, but hey, I'm not complaining.

Kathy was a-hooting and a-hollering as she put her light rod and line to the test. Expertly, she also was catching fish under another light 100 yards up the beach. With fish about the same size as mine, she took longer to land them, but her hook-up rate was as fast as mine with her jig and tail. By the way, I was using a suspended minnow lure that's shaped like a devil horse without the props. I cast, counted to five, and started a slow jerky retrieve that seemed to stay about 2 feet under water.

Eventually, my wrist started hurting so I reeled in, thank God, and

pedaled over to Kathy who was sitting in her kayak watching the show under her lights. Putting the rods in the holders we started the pedal back to the house. We were only 20+ yards or so off shore, and we do not have bow running lights on our kayaks (we do have a stern light) but we do have headlamps, so we have never felt any danger doing this.

The night was glorious as the wind had totally shut down, the stars were brilliant overhead, and we could hear our kayaks slicing through the water, some shore noises as people are celebrating the evening hours, and light kersplashes as mullet did their thing and jumped just for sheer joy. It's worth the trouble just to experience this sensory overload, but I always carry my rod and reel just in case. Reaching the house we beached the kayaks, washed equipment, cleaned fish, and had a glass of wine. Life is great. We got to bed about 11.

Next time, I'll tell you about bow shooting stingrays and the one that got even.

The CLAM RAKE AND I

The Clam Rake and I—
A Love Affair

My fingers tighten on the old wooden handle as her end snuggles affectionately against my neck and right ear. This way, I can hear signals the rake unerringly sends up the handle that she has found a clam. The rake and I are one, as we should be. After almost a year's absence, we're together again. Before starting to dig, I look east. The sky over the marsh is bright red with the morning sun warning of rain later that day, a good day for clam chowder or fried clam fritters. The old duck blind, dismantled by years of neglect, seems to hum stories of friends drinking coffee while watching a similar sunrise, perhaps over a large stool of decoys with a flock of canvasbacks starting an approach. The osprey nest silhouetted by the rising sun sits atop the blind; a mom is screaming at the fishing father, two starving chicks at her side. I chuckle at what I imagine she is telling him.

This smorgasbord of visual feasting goes on and on as far as I can see. That's why Kathy has a good camera. I, however, am here to clam. Tightly embracing the rake, I start to back up, pulling her tines behind me in the sand and knee-deep water at low tide.

For more than 100 years, this rake has plied clams from the Chesapeake Bay floor. Usually a member of the Callis clan is at the controls, but good friends Gene and Peg Callis let me borrow her to guide my efforts at clamming. This rake and I talk openly, with her doing most of the talking, me listening and thanking her when we have success.

There, a "clunk" comes up the handle. Knowing that it is likely a dead oyster or clam since the sound isn't quite right, I still stop and start to dig. The rake is always right, but I need a while to tune in. I stop, back the rake up a bit, then dig her tines in deeper, yielding a swishing sound as she cuts through the sand. I lift the rake, and up comes a large old oyster shell. I swear the rake says, "I told you so." I smile, toss the shell, and settle back into raking. It feels good to be home again.

The next time, the rake screams "clam" as it sends an unmistakable "screakkk" through the handle. Starting low, the sound climbs through a couple of octaves before ending (to me) in an unmistakable "k" sound. Feeling a little guilty at the excitement and glee that settles into my body, I back the rake up, dig the tines in deeper, feel the reassuring heft of a clam, and raise the handle. This lift lets the tines gently cradle the clam and hold it. I continue lifting until up comes the first clam of the season. I hold it and even talk to it a bit; then I toss it into the basket.

The basket is an old bushel basket floating in an old life preserver ring that resembles what must have been on the *Titanic*. The basket sits in about 12 inches of water but stays off the bottom so it drags nicely. A tether rope tied to the ring and to my belt lets the basket obediently trail along behind me.

Kathy, my love of about 40 years, is 50 yards away and I see her straighten and toss a clam into her basket. Soon, the clams and I will lose her attention as the scenic opportunities start to tilt the scale from clamming to photography. I am proud of her photographic work.

What is it about the smells around saltwater? I feel like my olfactory nerves are being assaulted with pleasurable scents. First, I notice the heavy salt-laden humidity in the air; then, the wetland smells move in with marsh flowers, decaying organic matter, and yes, even dead fish. After all, death is a part of life, even that part which makes life so valuable, so I don't mind the smell from a distance.

Enough of that—a cold chardonnay awaits us and the basket of clams. Back to "work."

A few more clams go to the basket, and I see three stingrays effortlessly gliding toward me, much like bald eagles soaring on a cliff over a river. They seem to exert no effort, yet their speed is impressive. Fighting the urge to attempt a walk across the water to escape, I realize a stingray would barb me only if he felt threatened. Thus, I hold my ground and wait.

Last year, Kathy and I both got into a small school, and they brushed our legs as they swam by. We and they did not panic, and we did not get barbed; it was quite a learning experience for me. Sure enough, these three see me and quickly swim out of sight, making all four of us happy.

Amongst all these ruminations, I am catching clams, and the basket is slowly filling. The big ones will be fritter or chowder ingredients while the smaller ones will never make it off the cleaning table. Raw, unwashed clams have an intense and satisfying salty flavor in a tender, easily swallowed body. They are really good with a cold beer.

I look up. Sure enough, Kathy's rake is stuck in the sand, and her basket is tied to the rake. Her camera has kidnapped her, and she is busily turning electrons into memory-generating images. She will show me those pictures tonight over the chardonnay and clams. We make a good team.

Right behind where she has tied up her basket is the bank that she and I fished this past summer for redfish, speckled trout, and bluefish from our kayaks. The first redfish, or red drum, hit her "zara spook" lure as she "walked" it across some very shallow water. A loud splash, lots of disturbed water, and giggles coming from her kayak were the show I got to experience. Later that summer evening, about 50 yards away, I saw a redfish bumping into the grasses, grass shrimp jumping, and the fins of the fish that were feeding on them. I paddled the kayak a bit closer and tossed a red fin up against the bank. I started a "walking the dog" trip across the top of the water. He swirled at the sound, his dorsal fin pointed directly toward my lure, and he rushed. He hit that lure like a linebacker hitting a quarterback, and the battle was on. We had redfish for dinner that night, as it was in the legal slot. I love skinny water fishing.

More clams go into the bucket. Taking a small break, I look south toward a lighthouse about a quarter mile away. All the land has been washed away, but the riprap remains, so the lighthouse seems to rise straight from the bottom of the bay. A closer examination shows that the sandbar that used to be land is still there. On the riprap, many types of shore birds show off their plumage, but I think the best is the orange-beaked Oyster Catcher that is a bit rare around here. There are always several around the lighthouse.

Enough clams; the rake and I are tired. I thank the rake—without her, I couldn't do this—stick her in the sand beside Kathy's, and tie my basket. Where is Kathy? I see her up by the beach lying on the sand, taking close-up shots of sand fiddlers (small crabs with very large front pinchers). I walk over, hug and thank her as well. Without her, this wouldn't be nearly as much fun.

Epilogue: That night we did have clam fritters, clam chowder, and cold chardonnay. Life is not good: it is GREAT.

Last Buffalo Turkey

When it started to rain, we moved to a double bull blind and set the decoys. Calling periodically worked, as three hens and four mature toms started in. Of course the lead hen came first, tilted her head, and looked straight into the blind through the open window as if she knew exactly what that green blob was. The turkey inside (me) moved his head just enough to see around her and watch those four mature fans now only fifty yards out. Immediately, I could see in her eyes that she had spotted me. She never putted but played relaxed a bit longer, turned, putted softly one time and slowly walked off. The toms folded fans, followed her and gobbled back at my calling but never came in.

Later, one of the largest beards I had ever seen followed them and just looked from more than 100 yards away. He was truly interested but very wary. Once he started in, exciting Jordan and me, but he never came closer than about 70 yards. That bird was a true champ.

Such was the end of a great morning. We had worked another mature bird at sunrise but bumped him accidentally and got to watch him trot off. Five mature birds, all interested, and we came close. I was hunting out of the Last Buffalo in Darrouzett, Texas, but hunting in Beaver County, Oklahoma, with three young friends of mine. Jordan Shearer, Josh Robertson and Jared Robertson, who run a quail/turkey hunting lodge chasing birds in Oklahoma and Texas. I have hunted out of the Last Buffalo several years and always get to work mature birds. It is a great destination for those of us who like to work older toms.

Back to the hunt, it was raining cats and dogs, and we were soaked. Back at the lodge, we dried our clothes, ate lunch, and took a snooze. About five PM, we decided that we couldn't kill birds in the lodge, so Jordan and I donned rain coats and went after them. After driving through mud for about 10 miles, we fishtailed through a gate (open, thank goodness) and stopped. It was pouring, but we got out and started to hunt. After about ½ mile, Jordan looked over a hill and saw a mature bird with 3 hens only 35 yards away. They also were miserably wet. I crawled over the hill (in mud of course) only to see they were now about 60 yards off which is a shot I will not take with my three-inch #5s through an old American Arms Side by Side. I like 20 to 30 yards.

I crawled back, we sized up the situation and decided to cross a creek, run up the other side and try to pop up on them before they crossed. As creeks have a tendency to rise when it rains, we went in over our boots, got to the other side, and started to crawl. About 150 yards later we got to a hill that would hide us, so we got to our feet and slogged another 50 yards. I crawled to a log and peeked over. The tom was standing 20 yards away. His countenance was very much like mine, "I'm tired of being soaking wet." The difference was that I had a shotgun. He had a 9-inch beard and close to 1-inch spurs. He weighed 16 pounds. This was not the most pristine hunt I ever had, but we earned that bird.

That night, we relived the hunt at the lodge and ate one of the best dinners (seriously) I have ever had. Going to bed about nine p.m., I slept great and thought of the hunt the next day. That county is a two-bird county, so I could take another and planned to be pretty selective on that last bird.

The next morning, we were sitting on a wheat field where we thought they would strut after that hard rain. Sure enough, we heard some gobbles, but they had hens with them and they decided an unseen bird is not nearly as valuable as those with them. Then, it started to thunder hard. We gathered the decoys and ran to the same double bull blind, set the decoys out front, and crawled inside, out of the pouring rain. Laying my diaphragm call on my knee, I grabbed a book, looked outside one last time, and started to read.

A few pages later, I caught movement out of the corner of my eye and looked to see two hens running by. The keystone cops routine kicked in. I slugged Jordan, who was also reading,

> The Last Buffalo is a restored old bank building that has made a great six-bedroom lodge with a large sitting area. The owners have decorated it nicely, complete with the only mounted Oscillated turkey, outside South Carolina, I have seen. On the walls are several large whitetail mounts, one of which I scored at approximately 180. They offer quail hunts for wild birds and turkey hunts for turkeys like myself. This year, they are going to offer whitetail hunts. I promise you some large bucks will be taken, as this is prime whitetail habitat.
>
> They have a professional cook who is awesome. We had great meals with dessert at each one. Even with all the crawling around, I put on weight, so the lodging is truly one of the hot spots, and the hunting is great.
>
> Their contact information is 806-624-1701 or www.thelastbuffalo.com

and looked around for my call. Finding it on the ground where it fell when I jumped, I plopped it in my mouth and yelped twice just as the same four gobblers stepped out. They heard the yelping, turned in my direction, and immediately went into full strut. That starts the comical "me first" routine where they race to the decoys. Of course, there wasn't a bit of space between their heads; I didn't know four heads could be so close together in a race. One was much bigger than the rest, and they all were mature birds, so he was the one I watched closely.

When they got to the decoy, he went into full strut, looked at the rest who dropped back, and he stepped forward for the last time. A 20-yard shot is not much for the same shotgun/shell combination and he dropped right there. The rest milled around and thought about flogging him, but they walked off. Jordan and I both believe that if the turkeys don't see people, they don't know what just happened, so we waited and watched. The birds didn't panic and just walked off following the two hens.

We went out and picked up our bird. He weighed 20 pounds, had a 9-1/2-inch beard, and more than 3/4-inch spurs. He was a nice Rio Grande bird, so his heft felt good on my shoulder as we walked out IN POURING RAIN. The next day dawned nice and sunny as I loaded in Jordan's truck for the drive home; doesn't it always happen that way?

> These are Rio Grande birds that are extremely healthy. We did not see a Jake the whole trip and we did see many mature birds. The 20 pounder had only ¾ inch spurs so he must have been a 2-year-old bird. That is heavy for a 2-year-old Rio, so I was pleased. While he was large, the lone tom that watched from 100 yards away had the thickest beard I have ever seen and HE'S STILL THERE. If you don't get him, I'm going to try again next year. Good hunting.

Coon in a Can

The big Bluetick hound had the raccoon by an ear, the Black and Tan had a tight grip on a foot, and I had a white five-gallon bucket into which I was to place the fighting coon (the official name is "raccoon" but henceforth, he shall be called a "coon"). Not being smart enough to say no, I dove in, placing the bucket on the coon and catching the Bluetick's head also. None of the four of us thought much of the situation so the fight really began…

Before going on, I'll describe how the coon, the dogs, and I got into this situation. A friend, John, said, "Sure, how about tonight?" when I asked about going with him on a coon hunt. The tragedy/comedy started. Hounds and bird dogs have always been a big part of my life, but I had never been coon hunting.

John, his friend Joe, the two dogs and I met at John's house at dark:30, drank a cup of coffee, and talked about dogs. All of us had been raised around hounds and had many stories we shared before going out.

The five of us (John, Joe, the Bluetick, the Black and Tan, and I) loaded into John's truck and started down the driveway. John stopped at his barn, grabbed a white 5-gallon bucket, like the kind restaurant pickles come in, and handed it to me. His response to my query as to "why?" was something like "you'll see." He was right.

On the way to the first hunt site, he told me they were having a meet the next weekend and needed a live coon for a treeing contest. I looked at him, the bucket whose purpose was then very clear, and said something intelligent like "you got to be kidding" (I was a bit more graphic). He said "nope."

Stopping at the first site, the five of us started into the woods. We humans had head lamps to help light the way, but somehow the dogs got by without those. Soon, the Black and Tan barked a deep, explosive, silence-shattering bark that said, "I smell one" (often called an "opening bark" by dog men). He went quickly from a "cold trail" bawl, through a "hot trail" chop, to a truly excited "running bark." The chase was on. Running through the woods I personally took out numerous spider webs with my head, got smacked across the face by several limbs, and tripped twice by tree

Photography by the Hero (Villain) of the story above: John Thornton of Stillwater, Oklahoma

roots, but I survived and loved it.

Before long, the hounds let out their distinctive quivering locating bawls which rolled over into a continuous chop telling us that they were "treed." Coons are smart, so sometimes they climb a tree and jump limb-to-limb until they can get down, so we couldn't be sure until we spotted it. We got there bruised, battered, bleeding and smiling. The two dogs were standing on their hind legs with front legs reaching up the tree, baying their hearts out. Our head lamps revealed two eyes reflecting light with an eerie glow. We could see eyes but not the coon at first. Then, his shape materialized as we continued watching.

John said something like "show time." Why did my heart start racing and my knees go wobbly? My time was about there. He and Joe caught the two hounds and I was left holding the bucket which, by the way, I carried the entire time we were running through the woods (then I knew why they were willing to take me). John said, "I'm going to call on my squaller and get that coon to come down. When he does, we'll turn the dogs loose to keep him occupied and you jump on him with that bucket."

I discussed his ancestry in reply, but he said something like "trust me, it'll be alright."

"Yeah!"

They caught and tied the dogs, John pulled out his trusty squaller and started to make a call that does sound like a coon fighting and sure enough, after a few minutes the branches started to shake and down came the coon to join in the fight that he was hearing. (Remember, he couldn't see us because we had our lights on him.) Excitement building, I was ready to go into the game, "Send me in, Coach."

The coon jumped to the ground, they let the dogs loose, and all hell broke out. Dogs were growling, coon was squalling, John and Joe were yelling, and I was shaking. It was time for my entrance, stage center. In I went and tried to put the can on the coon!

The big Bluetick had the coon by an ear, the Black and Tan had a tight grip on a foot, and I had the white 5-gallon bucket ready to plunge into the fray. Not being smart enough to run the other way, I dove in placing the bucket on the coon, catching the Bluetick's head also. None of the four of us (two dogs, one coon, and one nervous me) thought much of the situation, so the fight really began and the noise intensified to a level I didn't know was possible. John yelled, "Don't let the coon get out." I replied, "I'm afraid to let him go." As Jerry Clower sang, they needed to shoot the coon or me as we both needed relief. John was pulling his Black and Tan off the foot, Joe was trying to get the Bluetick's head out of the can without the coon attached, and I was hanging on as if my life were in danger. In fact, at that point I thought my life was about over and would have given myself a Catholic style blessing except my hands were both busy, so I just hung on.

It worked. Soon we had two excited dogs tied and the coon in the upside down bucket. Of course, I couldn't get up because the coon would come out. John calmly took the bucket lid and slid it underneath so the coon was trapped inside. He secured the lid and lifted the bucket with the now calm coon inside and unhurt. I, on the other hand, sat down caught my breath and licked my wounds which were all emotional and none physical. What an experience! John said, "Let's do it again, but this time, we'll just let the coon go, once we get him in the tree." Coon hides were valuable, but it wasn't coon season, so we were doing this for the fun of it. Frankly, it was fun, and I'd do it again in a heartbeat.

We got back to the truck, put the coon-containing bucket on the truck bed, and tried to do it again. At the end of the night, we got back to the truck and, somehow fittingly, the coon had gotten out of the bucket and escaped. The next night, they did it again, catching another coon for use at the meet. Oh yeah, we were college professors at Oklahoma State University, so we had classes the next day. Do the math, we hunted all night, taught classes most of the day, caught a few hours sleep that afternoon, and they did it again. I stayed home to recover.

Thanks John. That experience was a hoot.

> This story took place around 2000. The treeing contest involved a coon in a cage pulled up a tree. The dogs were judged on their treeing style and persistence. Although the coon was never hurt, most clubs no longer have such contests in respect for the animal. The hound culture still exists, and may it exist forever.

Fly Fishing the Colorado In Gore Canyon

Kathy and I spent a great day recently, fishing with Adam Groskin, a guide with Breckenridge Outfitters of Breckenridge, Colorado. We fished out of a wooden float boat that was a beauty to behold and quite seaworthy even in the class-three rapids of Gore Canyon on the Colorado River in west central Colorado. What a trip! We caught many brown trout and some large whitefish in the day's float. The biggest brown was around 18 or 19 inches and the average was around 15 inches for the 30 or so fish we caught. We would do this again and plan to do so soon.

Using 5- and 6-weight rods, we fished with thingmabobber indicators about 4 feet up from a large stonefly nymph (12 or 14) which was up from much smaller nymphs (sometimes one, sometimes two) such as a Trico of size 20 or so. The fish readily took all three and I'm constantly amazed at how a large trout can see and attack such a small obscure nymph in fast somewhat off-color water, but they did with a vengeance. All three took the browns and the really large fighting whitefish. Our biggest whitefish pushed 20 inches. The stonefly was the popular choice of the day for both the browns and the whitefish.

The most noteworthy experience occurred with my 6-weight, a relatively short leader (3X) and a large streamer such as a 10 woolly-bugger with a stinger hook. Adam told me to shorten my line to 20 feet or so and slam the shoreline close to the bank with the streamer. He said to use the rod to strip the line in maybe 6 feet and slam the shoreline again, concentrating of course on the pools. Entry splash was not a problem and, indeed may have helped get their attention as we were navigating class-three rapids close to the shoreline. I don't know the technical definition of a class-three rapid, but the waves were probably 3 feet from top to bottom and irregular, so you could not plan. My legs grabbed the platform with such an adhesion that it was impossible for me to fall out. This was somewhat like standing in a roller coaster and fishing with a fly rod, but it worked!

My very first cast was to the leading slick of a large bolder on the south side of the river. Immediately, a large brown flew from behind the bolder and struck the bugger with a vengeance. No short striking here as the stinger hook took care of that. Releasing that 16 incher, we went into the canyon where I had the most exciting fly fishing experience of my life. Hooking them was somewhat easy if I could get the fly against the shoreline, but landing them as we careened through the class-three water was another matter. Since we were going to release them anyway, losing a few didn't matter, and I was whooping and hollering the whole way.

Fast forward another 2 years and we did it again in the same boat with the same guide. We caught as many fish and had a great day. You need to try this.

I highly recommend Breckenridge Outfitters and Adam Groskin. Their website is www.BreckenridgeOutfitters.com.

Family and Turkey Hunting

"Families and Turkey Hunting Are Forever Wedded"

Obviously, I borrowed the spirit for that quote from Herman Melville, but I think the phrase says it well. There is no sport better designed than turkey hunting to bridge generations and keep families close, as these stories will show. One story to follow is retold but from a different perspective. I chose to let both versions remain.

At the time of this story, our son, Drew, and I have chased the Mott Bird for more than 2 years, compiling multiple encounters with the famed bird. We have one more chance that tests the combined brain power of two adult male humans against one mature male turkey. Thus far, the bird is winning, but so are we. Each such hunt brings Drew and me closer and lets us talk a lot.

Drew Turner, our younger son of 34, lives in Pueblo, Colorado, with his wife, Melissa, and their son Ethan, 3. I live in Stillwater, Oklahoma, where my wife and his mother, Kathy, and I raised him along with his brother, Travis. Travis is 35 and lives in Stockholm, Sweden. Kathy and I live to be outdoors and have raised our children to enjoy the same, so Drew and I get together twice a year to hunt. In the spring we migrate to Darouzett, Texas, to further develop an already tight father/son relationship and chase Rio Grande turkeys. Most of the time, Kathy comes along with her camera, and some years Melissa and Ethan tag along. When that occurs, there are more Turners in Darouzett than ever before in recorded history. In the fall, Kathy, Drew, and I move our camp to Colorado and chase elk and trout as an excuse to get together.

Darouzett is the home of The Last Buffalo, a quail-hunting lodge that doubles as a turkey-hunting lodge in the spring. The Buffalo is a restored bank building in the middle of Darouzett, but since I can throw a rock from The Buffalo to either end of town, traffic is not a problem. Each year we meet up with a very good friend, Wade Robertson, who along with his mom and dad runs the lodge. Wade is a very likeable Texan who got tired of playing baseball and the corporate world (he was a very good player at Oklahoma State University, I'm told). He moved back to his home town to do what he loves, hunt quail and turkeys.

Each year, when we first get back together, Wade drives up and stops—truck still running—in the middle of the wide street, door wide open. He walks over to us and shakes our hands. In the 30 minutes or so that it takes for us to renew our acquaintance, maybe two vehicles have to drive around his truck. Of course, those are also trucks, and the drivers wave and smile. I love this place.

Our first evening together there, Drew and I usually mix a drink and sit outside. The one blinking intersection light gives us the light we need to see (the closest stop light is 30 miles away). The wind is always blowing, so we get to hear the screen door blow open and slam shut. Naturally, it squeaks. The only activity is usually a bat or two working insects around the light and perhaps a wandering dog looking for an ear scratch. This trip, we retire early, as tomorrow morning we are going to chase turkeys and talk.

Early the next morning, Drew, Wade, and I down a cup of coffee, talk a bunch, and drive north to the Beaver River where Wade has spotted some turkeys, including a nice tom. Wade stops the truck, points west, and tells us there is a small makeshift blind at the end of "that grove of trees." Well, it's dark—we can't see his arm or the trees, but we head west.

It's a dark pre-dawn as we try to find our way without flashlights. We know the makeshift blind is up ahead somewhere, and we stumble over it in the dark. We place two hen decoys and one jake west of the blind. Drew and I settle down against two trees. He's shooting an old 1148 that my dad gave me and I gave him some 40 years later. I'm shooting a new black powder 12-gauge shotgun. The smoke pole is charged with 110 gr. black powder substitute, two wads, a wad cup, 110 gr. equivalent of #5 shot, all topped off with one more wad. This takes awhile to load but the pattern is great and deadly out to about 35 yards.

At first light, the gobbles start. There are about 3 gobblers announcing their presence and some faint hen talk waffling through the cottonwoods. I give faint tree talk back and they answer immediately. Through my face mask, I am smiling. Experiences like this are why we go to all the trouble.

During all this, Drew and I are sitting beside each other against the two trees. He and I both are able to slide the diaphragm call to the front of our mouths and whisper. We do as much talking as we do calling. This is catching-up time. Drew has often said, "Turkey hunting is not about killing turkeys." I agree, but I like to harvest one every once in a while.

After about half an hour of this communication, the birds fly down, still out of sight. We imagine they are strutting and trying to impress some accompanying hens. The gobbles go back and forth across the field as we vision that they're fanned and hot. Unfortunately, they also have hens, and no matter how seductive we might sound, hens in sight are better. We talk back and forth an hour, and they go silent. I can picture the hens starting to leave to go to nest. Drew and I also go silent, but we cluck and yelp softly every 15 minutes or so.

At almost 10 a.m. (note how close this is to the old lore about the hens leaving them about 10), we see a head stick up over a rise and look around. Its neck is stretched to the limit. It's a gobbler; we get ready, and he starts moving our way (about 100 yards out). As he gets closer, we can see it's a jake, but we keep up the soft flock talk. Every time I yelp or cluck, he flutters but doesn't fan, and he moves obliquely across the field. Since he never comes closer than about 50 yards, I'm thinking that our calling must have all the sex appeal of a slug—though there is another explanation, as you will see shortly.

He ducks under a fence and heads toward the river, but before he's out of sight, one lone hen starts toward us from about 50 yards south of where the jake first showed. She answers our soft flock talk with her own soft clucking and closes in until she is about 40 yards out. She drops behind a sand hill and disappears. A short time later, sand starts flying in the air, and we glimpse a wing periodically. She is dusting herself and putting on quite a show. We grin and keep talking; she's a great decoy.

Just out of curiosity, I let out a demanding series of yelps. She sticks her head up, looks at the decoys, yelps back, and comes on in. She stops about 15 yards out and starts scratching and feeding. Drew and I are silent now as she is simply too close to chance a call. After about 10 minutes, she stops, sticks her head up, looks over her shoulder and issues a series of loud evenly spaced clucks. There must have been 10 to 15 in the series. It sounded like an old hen using an assembly call, but they were clucks, not yelps. Drew and I turn slowly in the direction she's looking and there he is, more than 3 hours after we started calling. A mature bird, he walks in with no hesitation. He comes straight to the decoys, ignores the jake, and starts to strut. He is about 20 yards out.

Pulling the old "you shoot"—"no, you shoot" trick, we finally alert the bird and he runs off. Pulling our thumbs out of our anatomy, we laugh at this silly mistake and really appreciate the experience. We gather the decoys and head to meet Wade at the truck.

Pouring some coffee from a thermos into our cups, Drew, Wade, and I talk about the hunt. Is it possible that the jake was afraid to come in because he knew the "dominate bird" was right behind? When he fluttered, he wanted to strut and gobble, but after a beating or two he knew better and ducked out of sight. Perhaps he was the first scout. Also, the hen was the next scout sent in to check them. She knew big boy was right behind and she was protecting him. As it turned out he didn't need protection from these two keystone cop turkey hunters.

At about 11 a.m. and getting very warm, especially for two turkey hunters in camouflage outfits, Wade suggests that we go to the "Mott." Western Oklahoma and Texas is mainly short grass prairie with clumps of trees periodically. We call those clumps "motts," and I have no idea whether that's a word or not. It is to us. Wade drops us off about a half mile from the mott. The walk through the sage grass prairie is very hot and we are sweating profusely when we reach the trees. The large cottonwoods provide shade and cooling breezes that feel like an old friend welcoming us back. Each breeze brings in a sage grass smell that we really enjoy. We like this place. We are here to chase the "Mott Bird" again; but first some history is needed to set the stage.

We have hunted this mott 3 years in a row, and every year we have taken a bird or two. Last year, before Drew showed up, I hunted the mott and harvested a 3-year-old bird with a great beard and spurs. While picking up the decoys, I heard a loud gobble announce a newcomer. I sat down and started calling in the open (no decoys in place, no hide, and a dead bird at my feet). He marched right in like he owned the place (turned out he did) and went into strut about 10 yards from me. He was a mature 2- or 3-year-old bird with a sagging beard. I could not and would not shoot another so I just enjoyed the show. He spotted an ugly out of place pile of leaves (me), broke strut, gave a disgusted cluck and marched out of there. That was quite a sight and (I think) the beginning of a long relationship. The lore of the "mott bird" started.

Drew drove in that night and guess where we were in by mid-morning the next day? Sure enough, a gobbler answered our call fairly quickly, but this time, he was much more wary. He came in and stopped around 100 yards out. Since the sage grass is so tall, we could not see him, but we surely heard him. I tried to call him closer, but he tired of the game, marched into the prairie, and gobbled. This became a pattern that we were to experience again, and Wade was to experience with a client after we left.

Now, back to this year. The bird is year older (perhaps 4 years old).

The day before, I had gone to the mott in the afternoon and tried calling. About 3 PM, a bird answered in the sage brush, but he gobbled seldom and never in the same place. He circled about 180 degrees, gobbled a total of three times and disappeared. About 7 PM, I picked up the decoys and left, wondering if this could this be the same bird. Wade met me and took me to join Drew, Melissa, Ethan and Kathy at the Last Buffalo. Early in the morning, we went to the Beaver River (story above). Now we head back to the mott and this year.

We set up the decoys and back up against two large cottonwoods. We eat our oranges, nap a little, and talk a lot. We are calling every 15 minutes or so and we are trying to sound like a bunch of turkeys dusting in the midday. We sprinkle in some arguments just to make it interesting. He gobbles.

It's so quick that Drew and I look in different directions. He's out there in the sage brush but we're not sure where. We continue to call, nap a bit more, and chase shade around the trees for another hour or so. He gobbles again. He sounds off once each time and that's it, but this time we think we have him located.

We call another hour or so (do the math, that's about 3 hours total). Drew thinks he hears him again farther off so we relax a bit and talk some more. I move around the tree to better shade and take a significant nap. I wake, eat an apple, and had just picked up the slate and diaphragm to call when Drew shoots. This is startling and unexpected, so I jump a bit (actually I jump a lot and scatter my equipment everywhere). I look over at Drew and see one mature bird on the ground with three jakes beating the crap out of the dead bird, as they often do. When I move, they see me and run off.

Drew tells me he was actually almost lying down with his gun pointed toward the decoys and his eyes closed. He opened them and saw the mature bird march straight into the decoys with no hesitation. He shot, and the bird folded. We have taken the bird that we have hunted three times over two years, and we both feel excitement and remorse. He was quite a challenge.

The bird's beard is over 9 inches and the spurs stretch across the 1-inch marker on the tape. We estimate his weight field dressed at 22-24 lbs. This bird is special as through multiple encounters we had gotten to know him rather well. We will never forget this bird and the thrill of hunting him. This will give Drew and me something to relive many times over (hopefully) many more years around campfires. Probably the legend of the bird will grow a bit when Drew and his son Ethan are turkey hunting and remembering the old days. Perhaps the story will even last as long as Ethan taking his son or daughter out turkey hunting.

With all my heart, I hope so.

Today, vanity cost the Mott Bird his life. He bragged on himself many times to us, but he was wary. Finally, his pride (or some other drive) made him take the chance and move in. His life is over. However, I can only hope that perhaps I impress my adversaries as much as he did us and I hope that I can leave personal memories as strong and persevering as he did for us. We shall talk about him for a very long time.

We go back to the Buffalo, join up with the rest of the family, and take pictures. The day cannot get any better, so we clean the guns, put them up, mix a drink, and listen to screen doors blow open and shut.

Epilogue: I believe the Mott Bird had somehow learned that ugly piles of leaves (me) could hurt. Thus, for the next four encounters over two years (remember, Wade also experienced him with a client), he hung up out of range and learned that if he was patient, the hens would eventually come out to him. He simply had to announce his presence periodically. Furthermore, maybe he sat out there while Drew and I were calling, extended his neck, and surveyed the mott with his powerful eyesight. He spotted the decoys and tried to get them to come out. Persistence paid off for us and he just couldn't take it anymore. Gathering some jakes, he sent them in; but when they got close, he took over. He came right in not gobbling, ending the relationship. Sometimes, I think wildlife is much more wary than we think.

Please take your kids and/or grandkids hunting. Borrow a kid if you must.

Three generations of turkey hunters (left to right appear Drew, Ethan, the Mott Bird, and Wayne (me).

A Visit from Father Bear

Describing a dream is difficult even when it has a profound impact on you. Last night "Father Bear" visited me in my sleep and asked a very difficult question. I'd like to share that with you.

Getting to the heart of the dream and the message, a truck came by me on a back road. They were picking up bottles but had a gun in case they decided to deer hunt. They stopped and I asked, "Have you guys seen any bear around?" One said, "Father Bear lives on the next farm." I responded that I would like to hunt him; could they tell me more? "Sure, someone will be by to talk to you."

Later, a man came by who was faceless for some reason. I got the feeling he was tribal and was sent to me (could have been Marty or Bill, my best American Indian friends). This messenger said for me to follow him.

We went through the woods and came to an old house that I can still see. Inside the front door was a stairway going up. The messenger went to the back of the steps and looked around. I asked "why?" He responded, "just to be safe." I have no idea what that meant.

We went upstairs into a lightless room and the messenger disappeared. As my eyes adjusted to the dark, I spotted a black blob sitting in the middle of the floor. Somehow knowing immediately it was Father Bear, I sat on the bed and waited. Eventually, the bear said, "Why do you hunt me?"

I woke up unafraid and not nervous but profoundly impacted. I had started hunting bear and lion and never quite felt comfortable doing it. I would love to kill a bear and a lion and have promised I would eat as much as possible of both; but I'd be lying if I said meat was my goal. This is an ego thing to have a bear rug and a lion mount in my game room.

What does all this mean? Both of the American Indian friends mentioned above immediately said, "You should not hunt bear, lion, or any other predator, for it could be very bad medicine." I have not hunted them since. Weird eh?

A Horse Packing Trip Through
The Sierra Madres of Mexico

A hole dug in the stream bank and clothes lying to dry portrayed the presence of Tarahumaras, an Indian tribe in the Sierra Madres of Northern Mexico, as we rode by on horseback. Likely, they heard and/or saw us coming and disappeared into the surrounding forest where they sat and watched. They are a fascinating people who are very shy after being persecuted for so many years. Other than some of the guides helping us, this is our first encounter with the Tarahumaras but there would be many more as we grew to appreciate these people and their culture.

We were on a 4-day horseback ride through and around Copper Canyon, Mexico. It was beautiful country, and getting to know the Tarahumaras was an eye-opening educational experience. We were led by Gary Ziegler of Westcliffe, Colorado. There were nine of us in the party, all from Colorado or with ties to Colorado. We had flown into Los Mochis, Mexico, where we took a van ride to El Fuerte. There we stayed one night in a very old restored hotel (Rio Vista Hotel) before boarding the famous Copper Canyon train. The hot water shower in the hotel trickled at best but the scenery and friendship were great. The tortilla soup and margaritas put an exclamation point at the end of a great day. The train ride itself was beautiful, but we shall describe the horse pack trip.

We came in at the rim of the Canyon, Posada de Baranca, Mexico, where we stayed with a Mexican family for two nights. Lola, the mother, was apparent head of the lodge. Their home was the kitchen and dining room; bedrooms were in a motel-looking structure outside. Guerillermo, our head cowboy and Lola's husband, could shoe a horse in the wilds faster and apparently better than anyone I'd ever seen. The horses never complained so he must have been good. The meals were absolutely fantastic as were the margaritas. We rode one day just outside of town looking down into Copper Canyon and other side canyons. I don't know how to compare the width and depth of Copper Canyon to our Grand Canyon, but I do know I was just as impressed with one as the other. The canyon alone is worth the trip, but we were there to explore deeply into the canyon on horseback and camp each night.

On the first day, we mounted our already saddled horses and rode out of the village. Within 15 minutes, I felt as if I had left the world behind and was entering a new one. Up and down rock- and boulder-strewn paths with the click of horseshoes on rock, and the frightening sound of shoes periodically sliding over rock took some getting used to on this first day. No one fell, but the ride was very exciting. Soon we encountered the laundry scene described in the first paragraph above.

Lunch the first day was in a beautiful box canyon with a cave, ruins, and burnt walls from many fires staring down on us. Across the canyon, a more recently blackened rock gave away the presence of fire. We focused our binoculars on the spot and saw a much more recent lean-to where "they" might have spent the night as their goats grazed in the valley below. I felt strongly as if someone were watching us, so I took a side trip to the dry creek bed below to check on tracks, and sure enough there were goat tracks, fresh goat poop, and sandals. Indians had been here recently and perhaps were still watching us from high above. COOL!

Now, we are a few miles deeper in the canyon, camped along a stream, taking rum straight, and solving the world's problems. The sun was disappearing behind a small mountain due west. Theresa, our chef, could better prepare meals for 20 cowboys and gringos over a campfire than we could at home. Each evening we ate like royalty (after the rum and margaritas). Life was great, and we were so fortunate to be able to do this. We slept well in the clean environment that cooled much more than I expected. My down bag felt great.

Overnight, the lead donkey was tied to keep him and the others from wandering, but not being free was very frustrating to him. He let us know by braying all night—a sound that has much more inflection than I remembered. Kathy and I slept through most of it, but some of our group had trouble. The bray was a long mournful lonely call that he issued every hour or so. Actually, I loved it.

On the second day, we spent as much time walking our horses as we did riding. The terrain was very steep and rocky for much of the day and although the horses were extremely sure-footed, we felt we could get out of the way better if we walked. While walking, I stepped on a boulder and it rolled. Switching hands on the lead rope quickly, I stuck my right arm down and the terrain was so steep, I was able to use my arm as a prop and pop right back up. Only I was impressed with my agility as the rest were laughing too hard to be impressed. Later, I'm on horseback when my horse "Socks" slipped to her knees, quickly regaining her footing. Neither of us fell, but it sure got my attention.

Lunch on the second day was on a stream below a rock cliff with colorful drawings and a rock shed with three shorthorn bulls or steers. They played king of the mountain and never left the rock shed. We didn't argue, but I would have liked to have seen the drawings. We had a great lunch with sandwich meat and several local cheeses. I walked from about 11 AM to 3 PM when we reached our next campsite, giving my horse's leg a rest, and burning some calories myself. In four days of eating and drinking like an orgy-bound Roman, I did not put on any weight.

Our campsite was a lovely long sand bar high over the Rio Oteros which flows crystal clear and cold. I love flowing water, so I walked the river exploring for life. There were lots of minnows, a nymph that looked much like our stone fly nymph, some tadpoles, and an extremely successful underwater spider that was hell on local flies. I saw no fish bigger than about 2 inches.

We bathed in the cold water and found it very invigorating. Kathy made me keep my wading pants on, stealing a bit from the moment but great none the less. As I wrote this, I was sitting on a rock drying, warming, and anxiously awaiting the margaritas. I did fall asleep but was on time for happy hour. We ate a very good meal of chili relinos and the ever-present frijoles. I assure you that if you ever take this trip, constipation will not be a problem. Once again, we slept extremely well.

Our third day in the saddle was over terrain very rough and steep, so we walked much of the day. We had a great lunch on the banks of another stream and then walked into the village I describe below. I found this village experience quite moving and spiritual. I doubt if I can convey just how much so, but I shall try. I would have ridden those four days just to experience this village and the local Indians. We were high on a hill camped with a view of the valley. A Jesuit-inspired church of maybe 20 square meters sat behind me as I sat on a rock taking it all in. I started the write-up EARLY the next morning:

A pastoral scene of incredible beauty is stretched out below as if on a stretched canvas. My breath was literally drawn out of my body as the scene and its message sank into my thick skull. This message is profound, please read on. The sky is brightening to the east as day is winning its battle with night, promising a bright sunrise in a few minutes. A clear creek starts in the far right, meandering to the left (north) before turning 90 degrees east and curving out of sight through peach trees starting to bloom. Three Yorkshire-looking pigs are feeding under the peach trees. I'm sure their lineage is unimportant to the Tarahumaras but they look like Yorkshires. The pigs feed into the cornfield rooting around for goodies and fertilizing the field in the process.

The Mexican and Indians are gathering the horses and burros for the day's ride and are corralling them in the northwest corner of the field. Cow bells on some of the burros, the braying of the lead burro, and the birds singing add to the full-dimensional sensory experience. Of course, they also added their contribution to the fertilization.

The cornfield of maybe 10 acres is 2/3 plowed, but an Indian is using a horse and a single-point plow to complete the job. The man and the horse struggle with the job that a tractor could complete in 3 to 4 hours but will take the horse and man probably a week.

A flock of chickens is happily following the man and horse scratching for bugs in the newly plowed land and fertilizing the field in the process. Several roosters are in the flock challenging each other, but the hens are content to just scratch. This is sustainable agriculture at its best.

This Indian village of maybe four or five families is so remote that no tractor could get here, and they wouldn't use it anyway since there is no fuel. Anything they consume is grown here and seed is recovered for next year as there is no access to outside commerce. There is no irrigation carrying silt to the creek and no fertilization other than what pigs, horses, and chickens produce naturally. Of course, there is no electricity or pumped water (other than gravity-fed springs to some of the houses)—what you see is what you get. I tried hard to see what pollution they might produce, and so help me, I could find none.

Yesterday, I watched a couple of kids swimming and frolicking in the stream. Some were on horses bareback and some simply were wading. I didn't see any soap.

The summers can be very hot, so up on the hillsides another 300 meters or so, are a ring of perhaps five houses looking down on the valley.

When it's hot, the people move up; when it's cold, they move down. The strategy is effective as not only is it cooler up high, the wind blows much more helping with the cooling.

I have never seen a place so much at home with the world. These folks literally live with the land rather than yanking an existence from it. The yield of the cornfield is very low due to the 20 inches of rain and no fertilization, but it is enough for them. The corn fodder is feed and roughage for the horses. Everything grown here is consumed and everything consumed is grown here. We talk about our helping indigenous people from around the world learn to live as we do. Perhaps we have the teacher and student backwards. We could learn so much from them.

The Jesuits left their impact in that a very modest but extremely well cared for chapel sits in the middle of the village overlooking the valley described above. In a very somber ceremony of sorts, they allowed us to enter the chapel singularly or in twos. Kathy and I were profoundly affected by this and felt closer to God than we do in our own church back home.

There is an ephemeral and spiritual feeling about this knoll upon which we perch. Gary just described their religion as animism, and I agree in that every element is tied together. Their farming, their religion, and their families are all one and the same. They are intertwined so tightly that it would be impossible to talk about one without the other. Wakan Tanka is alive and well, flowing through their entire culture.

We mounted and started the last day's ride or actually walk out. The going was very steep and we walked for a couple of hours which was exhausting and exhilarating. Then we alternated walking and riding until we reached our lunch site at the base of a cliff dwelling. We hiked up to the dwelling where we lunched on sardines, oysters, squid, and octopus (great food) and Gary told us more about the geology and history of the area.

After lunch, we climbed back on horseback and rode the rest of the way out. The horses and the people all knew we were headed toward hay and hot showers, so the ride was quick. We arrived at the lodge where we all showered and had margaritas again. Dinner was awesome as was the entertainment which was provided by a local and his son. The boy was quite talented.

Next day we rode the train back to El Fuerte where we stayed in the same hotel, once again a great experience. Then, we took a van ride to Las Mochis where we boarded a plane and flew back to the US.

The experience was unbelievable and great. As you can tell from the write-up, I was tremendously impressed with the Tarahumaras and their ability to live with the land. That lesson is still sinking into my western skull.

Letter to Ethan (grandson)

Dear Ethan [our recently born grandson],

Today is Christmas Day 2003, your first. We have several gifts for you and are eagerly awaiting your and your parents' arrival this afternoon. This letter is one of your gifts, but it will mean little to you until about your 16th birthday. I hope you keep it until then.

You came into this world a little early and had quite a battle to catch up. That you have done, and we are extremely proud. You are quite a fighter; I hope you always stay that way. Your grandmother and I truly love you (as do your parents).

I won't waste time talking about all the virtues that I hope you develop, but they include love for your family, courage, honesty under all trials, appreciation for the natural beautiful world around you, and the deep satisfaction of doing a good job. If by 16, you don't have those qualities, we have done a hell of a poor job raising your dad and he has done a hell of a poor job raising you. We will be trying.

I am leaving a few things for you. First, I leave you some fly rods and rifles. I leave you a 30-30 that Ed (Ruark) and I wrangled a half day for with a pawnbroker in Christiansburg, Va., around 1970. It is a Model 94 pre-1964 Winchester. It is worth a little in money but a lot in memories. I have harvested many deer with that rifle; your dad took his first and second with it. I hope you follow suit. It means a lot to him and me. I leave you at least one of my fly rods (I have to be conservative in case there are more of you). It will be a 5-piece, 6-weight Winston with Bill Mashburn's name on it. Bill is a good friend who gave me that rod. Now age forces him to talk about fishing more than actually fishing, but Bill is always with me when I use that rod. I hope the feeling comes through with it. I leave you a Ruger 9-mm handgun that I seldom use or see. You will want to keep that someday to protect your family. I strongly believe the role of protector comes with being a dad. Finally, I leave you a host of collectibles—my life-long collection of flies which is more impressive than my skill in using them and my duck stamp collection which is complete. Barbara Bramwell's dad who died some years ago started this stamp collection. I have kept it going since then.

Enough of that. Now I will tell you the important things we leave you. My dad passed them on to me, your grandmother and I have already passed them on to your dad, and we all will pass them on to you.

I leave you cold gray days with the predawn light illuminating your decoys. The whistle of duck wings overhead, the sounds of waves breaking against the shoreline giving your decoys the necessary movement, the sounds of your dad and me calling and pleading, and finally the welcome feel of a shotgun as you harvest. I leave you the appreciation of that life you just took, as we clean and prepare the wildfowl. The eating is great, but way behind the memories the eating reawakens.

I leave you the self-satisfaction of pulling in your crab pots and harvesting the delicious life we call a blue crab. The feeling as you prepare and eat the crabs with close friends and family is the culmination and most important part. Always stay close to good friends. If they ever lie or cheat, drop them, as they aren't friends. Your grandmother, dad and I have many good friends; I hope you do also.

The exhilarating feeling that you get when that large cutthroat (bow, brookie, salmon, etc.) takes your fly after you have worked him/her for an hour is one of my gifts that is extremely important. You spotted him feeding, you had him hit two different flies on two different passes, and finally he strikes a large hopper with a vengeance. Half an hour later, you land that 18-inch cutthroat, kiss him, take a picture or two, revive him carefully, and release him to do it all over again. Trout don't live in ugly places, so you stop to enjoy the moment, to look around at the beauty, and to smile in just sheer appreciation for being there. At that point, it's all right if you shed a tear remembering all the trips you, your dad, mother, grandmother, and I will have taken. I assure you that we do and will.

I leave you the warm summer evening as you arrive at the pond to harvest some frogs. You turn off the radio (country music is a must) and open the door. The warmth and humidity assault your senses immediately, as do the loud shrill calling of cicadas and the persistent humming of bugs. The warm feeling as you step into the water and begin the tour of the pond coupled with the thrill of seeing your first frog frozen in the light. The ultimate harvest is like icing on a cake. The excitement when you spot a deer on the shoreline or a coon looking for craw dads is important to the total experience, but the most important is the fried taters and frog legs you enjoy that night with your good friends and family. After all equipment is put up and you do not have to drive anywhere, it is all right and even important to share a drink of good whiskey or wine with those good friends. Drink responsibly, please; we have many good memories to generate.

Your grandmother and I leave you the "little pleasures" of being outdoors. The warmth of the sunshine, finding a spot out of the wind, and the wisdom to know you must stop and soak up that pleasure are some of those. The coolness and taste of a spring as you stop your horse after a long dusty ride and both of you enjoy the moment as you develop a closeness is another. The sheer beauty of a spectacular sunrise/set as you watch that buck tend his scrape with your heart literally pounding a hole in your chest as you wait is one of the more thrilling pleasures. There are thousands of others that we intend to introduce to you as you grow.

We leave you a cold winter evening with a fire in the fireplace, close family/friends to talk to, the sound of sleet hitting the window and the sight of big snow flakes as they pile up outside. If you're lucky, those close friends will include at least one hunting dog who, I promise, will always be among your truest friends.

All these and much more we leave you, Ethan. Hopefully, we will live long enough to have you read this; but if we don't, please take time to remember those great experiences that we will have. Today, your great-grandmother, grandmother, mother, dad and I will lift a glass to you. You will not know it, but you will find out 16 years from now when you read this. We all truly love you. MERRY CHRISTMAS!

December 25, 2003
The idea for this letter was taken from a famous
similar letter written by Corey Ford.

Antelope Blind

A slight breeze awakens the sleeping windmill. The vanes slowly turn, delivering a fresh supply of water into the stock tank. The almost eerie creaking sounds of the mechanism awaken the dozing ravens who start to mutter moans and trills that are quite soft and musical. I have never heard those sounds below but I must admit I had never before sat in 100°F heat beneath sleeping ravens.

The heat inside the blind is almost unbearable, especially since my black clothing is chosen to hide inside the black interior of the blind and they are extremely hot. The clothing starts to come off. So much is comical, but not worth writing about.

My gosh, the stock tank is full of cooling water with only a bit of algae, hmmm. Hopefully, the antelope think that is inviting, so I continue to sit inside, roasting in the heat, writing, drinking water, and thanking God I do this. I sit, and when bored, I sit some more. All day in this heat.

NEXT DAY'S HEADLINES IN COLORADO SPRINGS NEWSPAPER: 72-year-old grandfather arrested for skinny dipping in a local stock tank. Psychological evaluation ordered.

Best Friends

You were my best friends, always my first choice for hunting-fishing-camping-frog gigging trips. That you were also my sons added to the fun and fulfillment. Frankly, they have not been the same trips since you "grew up," but you do still share some with me.

Travis, you always were the campfire chief, from building the fire pit and fire to tending it. You always did your job well, and you also loved to limb line for bream. A book frequently accompanied you on our duck and deer sits, and I always felt close to you as we sat, me waiting and you reading. Thanks.

Drew, you were always willing to go fishing or sit in a deer tree stand with me. I always loved it when you got chilly and sat in my lap. Hugging you and talking may have caused us to not kill deer, but it sure made for a great deer hunt. You once told me you could not imagine a better childhood. I cried with joy inside. Thanks.

Now, you're both off on your own and I am left with incredible memories. If I were to have written my ideal sons' description, you both would exceed that but in different ways. Thanks for sharing your youth with me.

Now, you have the baton.

Dad

A Great Snow Goose Hunt

Sometimes the geese came in small flocks that approached the decoys, lowered feet, spread wings, and dropped into the spread giving us easy shots. These were often younger snow geese retaining much of their blue-grey plumage.

Sometimes they came in huge flocks flying very high. These often soared and looked through four or five passes before either dropping close enough for a shot or flying on to other pastures. These were often led by mature solid white snow geese that had seen many spreads before and had lost buddies, they were wary.

Usually, however, they came in medium-sized flocks of mixed plumage (white and blue). They would often make a number of passes but these normally could be seduced into our stool by the calling from a number of speakers spread around the 1,700 or so decoys we had and expert calling by our host Chris Schiller.

My family and I (Kathy Turner, wife and photographer; Drew Turner, son; Ethan Turner, grandson; and Annie, yellow lab and great friend) were hunting in northeastern Colorado with Chris Schiller, renowned classical artist who also is a farmer and our host for a snow goose hunt. Chris is a great blind companion and an unbelievable cook inside the blind whose calling sounds more like a goose than a goose does. Breakfast burritos or biscuits and gravy are awesome during lull periods and the espresso (seriously) really capped off the experience, but we were there to snow goose hunt, and what a hunt we had.

We hunted on March 18, 2010, a blue bird day. The morning started with a glorious red sunrise as a cold front and snow were coming in. The temperature was around 30°F but it quickly warmed to around 60°F with a strong south wind that turned about midday to a north wind foretelling of weather to come. The winds gave the single pedestal decoys motion that must have looked very attractive from above for almost immediately after legal shooting time as I was parking the truck and walking back to the blind, Drew and Chris both shot geese. The day was starting.

I got in the blind, looked around at my family and Chris and thought "it just doesn't get any better," but it did. For the next two hours we worked geese, shot many, and made the observations I noted at the beginning of this story. This was the first hunt for our 6-year-old grandson, and his wide eyes made the whole experience special. He and Annie would together retrieve the geese, but Annie usually won the race. Then, a lull occurred and Chris demonstrated his culinary skills which are considerable. A little later several more flights came through.

We ended the day shortly after noon for the front was seriously coming. The five of us walked out with our 17 snow geese, loaded into the truck, and drove to Chris's house where we looked at and bought some sculpture. We bought two bronzed bookends (a beautiful symmetric whitetail buck on one end and a regal asymmetric mule deer on the other), but we drooled over some of the other sculptures he had in his home. The artwork was worth the trip but the snow goose hunt was AWESOME. Our grandson summarized it all by asking, "When are we going to go again?"

We quickly loaded into the truck for the 4-hour drive to our home in Fairplay, Colorado, while Drew and Ethan loaded into their SUV for a 4-hour drive to their home in Pueblo, Colorado. The temperature was still 59°F when we left Chris's driveway but steadily dropped during the drive to 29°F when Kathy, Annie, and I pulled into our driveway. Only a few snow patches were on the ground. Next morning when we got up, there were about 6 inches of new snow that quickly grew to 12 inches before it stopped the next day. As I wrote this, our thermometer read 5 below zero. Clear blue skies, 12 inches of snow, and a brilliant alpine glow on the mountains behind us made for a special start. Living in Colorado is GREAT.

For the hunt, I was shooting an American Arms 12 ga side-by-side with improved and modified chokes. Since a side-by-side kicks like a mule, I limited myself to 3-inch BBs, but would have done better with 3½-inch shells. Drew was shooting an old 1148 with an open cylinder. Since it can't handle 3-inch shells, he was limited to 2¾-inch Hevi-Shot. Chris had the ideal gun for this hunt in a Winchester SX2 semi-automatic into which he loaded four 3½-inch 2 shot. He was deadly with that gun. If you're serious about snow goose hunting, I would recommend 3½-inch shells, and a semi helps considerably with the recoil.

A Triple Header

Introduction

It's 30°F outside and snowing in Fairplay, Colorado, the wood stove is roaring as the word processor and I start our sojourn and story. It's hard to believe that tomorrow morning I leave for a turkey hunt in western Colorado (April 12, 2011) at High Lonesome Lodge, a beautiful lodge set in the mountain foothills considerably lower and warmer than we are at 10,500 feet. Lodge personnel tell me the turkeys are gobbling and the trout are feeding. What a trip—we are hunting in the mornings and fishing for large trout in the afternoon; life is good. The story of this hunt is titled **"Fins and Spurs."**

This is part of a three-trip odyssey starting with High Lonesome after Merriam turkeys. Then, we take a trip to Virginia to run our crab pots and fish on the Chesapeake Bay. On the way, I will hunt Rio Grande turkeys in the Oklahoma panhandle with some good friends. The story of this hunt with my son and two very good friends will be titled **"Spurs and Longbows."**

The final leg, an Eastern bird hunt, will occur at a friend's farm in Virginia. This will be done with the shotgun. This third hunt will be titled **"Spurs, Crabs, and Stripers."** So, stick with me; these will be great hunts.

Fins and Spurs

For an hour I have watched this mature tom strut no more than 30 yards from me, sometimes as close as 20, but I can't shoot. His mature fan is shinning in the backlit sunlight, showing it is very healthy and full, and I can see him shudder all over as he spits and drums really adding to the color, but those darn willows are between him and me so that at times he disappears, then he shows himself, but he never gives me a shot. Yes, I could have sent a load of #4s or 5s bursting through the brush, as I shoot a double barrel American Arms shotgun with 5s in the full choke barrel and 4s in the modified, but I am not about to wound a bird like this; it's all or nothing.

I was hunting with my friend Todd. The tom had answered Todd's and my call more than a quarter mile away. Todd being the primary caller sets up immediately, but I being the only one armed head toward the bird

High Lonesome Lodge lies in western Colorado at about 7000 feet just outside the town of Debeque. Only 30 minutes away is Grand Junction, Colorado, which has excellent flight service provided by several major airlines. The lodge is on the shore of a stream which provides water for the many trout ponds scattered throughout the valley and irrigation for the hay fields which are absolutely over run with mule deer and elk. On the way to and from our hunts, we would easily see several hundred mule deer and maybe 100 elk. A tom turkey fanned in his full glory with as many as 20 hens would round out the wildlife experienced. It is truly a wildlife Mecca. As the story relates, we saw the largest black bear track I have ever seen.

The ranch offers great bird hunting for chukars, pheasants, and quail as well as large game. Large game include elk, mule deer, Merriam turkeys, bears, and mountain lions. Being the home of perhaps the largest concentration of lions in Colorado, the ranch is presently collaring and protecting mountain lions as part of a research study, so lion hunting is on hold. It's fun being there knowing that mountain lions are likely watching you.

As you will see in the story, fishing is excellent for large trout that are wary, but the fishing is not difficult or especially technical. The fish are healthy and fat. You go there primarily for the turkey hunt, but find spectacular fishing.

My wife, Kathy, enjoyed working on her photography while I was turkey hunting and joined me for the fishing the rest of the time. With her normal acumen, she caught the largest trout.

Finally, "you are never far from your next excellent meal." The food is the best I have ever had in a lodge, and it is prepared/presented exquisitely. The staff is very friendly, and they work hard to make you feel like family, which indeed you are.

For more information visit www.highmountainlodge.com or call them 970-283-9410.

---

30 yards and sit down. I can see the hillside he likely will come down but those willows! I stand to go forward to the hillside, but there he is in his full glory. I sit back down and watch. Surely, he will come out from behind the willows enough for a shot. At 30 yards, my American lays a deadly pattern so all he has to do is show, but he never does.

Todd calls for half an hour or so with normal yelps, clucks, and cuts. The bird gobbles multiple times, so he is hot. Todd starts to cackle, cutting off his gobble, and that turns the bird on, so he gobbles more but doesn't move. Todd pulls a final trick, he gobbles which flat out turns this bird upside down. The bird gobbles four times in a row once, but he never moves from behind the willows.

After an hour, he folds fan and starts up the hill so I put in my diaphragm call and start cackling, hoping that the closer sound would make him think the hen is coming, and again, the bird gobbles so loud and often that I think I feel the ground vibrate. He turns, starts back down, but stops at the willows again. We can't budge him.

Todd and I crawl out and listen for him to go to roost. He gobbles a couple more times and then silence so we know where he is. Tomorrow!

The next morning well before daylight, we are on the edge of a clearing where we think he might fly down, and we set up a little uphill thinking he would like that. Todd does a few tree calls and the bird responds as in the night before. At early light, the two hens fly down and the big boy obediently follows, but the wise hens fly farther down the field and land about 100 yards from us. The game begins again.

Todd and I call softly like a flock, we later cackle, and pull every trick we can think of except for moving in on them. Exactly like the night before, I can see him the whole time but can't shoot as it's too far, but this time he is often in the open looking up at us. The hens slowly start to move him away (two in the hand are worth more than two in the bush?) but periodically, he comes back to the field, thunders a gobble or two, fans and turns so we can see every angle. Finally he folds and trots after the hens.

Giving each other high fives, Todd and I talk excitedly about how much fun that is. I honestly would rather that happen that have him fly and land in my lap. We'll get him or another, now that we know his patterns. We have now worked that bird 2 hours with him in sight the whole time. (Yeah, I know, but at least Todd is a good caller. Most birds would now be a jelly head.)

We have three other toms patterned so we spend the rest of the morning checking them. Gobbling is not good as they are with hens and the hens want to protect them. We spot one flock and attempt to stalk within calling distance but they disappear. Perhaps they spot us as we start the stalk and run quietly into cover; anyway, they are gone.

On the way back to the truck, a road grader is working, and he needs to back up. He puts it in reverse, activating the "beep, beep, beep" that is a safety measure. Immediately a bird on the mountain about ½ mile away complains with a loud gobble that we can just hear. In fact, we disagree as to exactly where he is, but we take a bearing in between the two directions and take off up the mountain. We get quickly into scrub cedar with lots of deer, elk, and turkey sign but we can't raise a gobble. Continuing, we spot a huge bear track. I wear a size 11 hunting boot and his rear foot was at least one inch longer than my boot. Measuring the front foot, Todd estimates the bear would square over 7 feet. Anyone want a large bear? High Lonesome offers bear hunts.

Continuing up an old four-track road, we start seeing lots of turkey sign. We try a coyote call and loud yelps but no response. The poop is fresh and some of it is emitted by a large gobbler so we hunker down and start to still hunt. Sure enough, we see several hens that are wary but not spooked. Sitting, we wait until he steps from behind a cedar. A load of #5s interrupts his journey and we have our turkey. He is a nice mature bird with 8¼ inch beard and a beautiful full fan. His spurs are short as often happens in Merriams up here; in fact, the ranch has several mature birds mounted with "no spurs" other than a rough spot where the spur should be.

Step one is accomplished; I have my Merriam. On to the next step, Rio Grande birds in Oklahoma and Texas.

That afternoon, my wife Kathy joins me and we go fishing with Jessie, also a friend and fishing guide. We are after large trout with small flies. I am using a 5 wt rod and Kathy is using a 6 wt; Jessie is trying to get us into fish, but I insist he bring a rod for himself which he does, though he is truly interested in our catching fish, and he is delightful company.

We stop at one of their trout ponds and start fishing. My 5 wt is tossing a 16 Bead Head Nymph with a dropper of a smaller Caddis nymph flashing green and no strike indicator. Kathy is using a 14 olive woolly bugger. We start catching fish immediately and catch them all afternoon. I am catching mostly rainbows, and she is catching large browns that really seem to like the woolly bugger. We probably catch 25 large trout between us before dinner calls. I would estimate the average size trout as 17 inches, but we caught some as small as 13 inches and several 20 inches or more.

The next two days we fish with Jessie at different ponds using different combinations of flies, all with the same result, lots of big trout. We catch cutthroats, rainbows, brookies, and cutbows. We never measure the fish but estimate, conservatively, that about 10 over the two days are 20 inches plus and that we catch somewhere around 50 trout, all released to be there when you go.

Between catching large fish, we eat exceptionally well. Aunt Linda, the breakfast chef, fixes old-fashioned meals with everything homemade. Chef Jordan continues to wow us with great lunches and dinners all exquisitely prepared and presented. The wines and unique deserts really add to the experience. This is a turkey hunting and fishing trip that I shall do again, and Kathy says she is coming with me. I hope you try it also.

After a long drive from Colorado to the Oklahoma panhandle, picking up Ed Ruark in Amarillo and Ed Hastain in Colorado, we arrive at the Last Buffalo to start the

> **At the Last Buffalo Lodge,** quail is the normal quarry but Josh, Jared, and Jordan also offer turkey, deer, and duck hunts. I have hunted with Josh and Jared's dad for many years, but now the boys run the operation. It is one of my favorite places, and we are here to turkey hunt.

second hunt on my triple header. Already a mature Merriam turkey has fallen and some large trout have been caught at High Lonesome Ranch in western Colorado. Now, a Rio Grande bird is the goal.

Spurs and Longbows

For those of you who might not know, a Merriam has pure white for the tips of the tail feathers, a Rio has tan, and an Eastern, the next quarry, is dark brown. I find there are subtle differences in hunting the three different species, but the similarities are very strong. Since I kill the Merriam early in the season with snow on the ground, the Rio mid-way through the season, and the Eastern toward the end of the season, the differences could result from the presence or lack of hens.

"Oh, shucks, they are spooked." The first afternoon, on the way to a preset blind with my longbow, I spook a bunch of mature gobblers but not badly. They walk off so I think they will come back. With my three decoys set 15 yards in front of the blind, I climb in and open (I thought) the necessary windows. To cover my drawing the bow, I like to leave the back side closed (guess which direction he comes from), but I have great shooting lanes in front.

He answers my calls fairly quickly and starts in from behind me, so all I can see is him through a small opening in the window. He struts to within ten yards of the blind and lies behind a log waiting for something while I am scrambling inside the blind trying to figure how to shoot. I manage to open the window a bit more, stick the arrow through the window, draw and shoot.

Now, I have shot turkeys out of a blind with a primitive bow before, but always I had an open window. If you ever have a chance to do what I just described, don't! As I released, of course the limbs flew forward hitting the blind and sending almost 700 grains of arrow off into space with a loud resonating noise. The bird just looked around like "what was that?" Not being a quick learner, I decided to try again but farther inside the blind with the same result for both the arrow and the bird. By now, I am shaking like a leaf.

The bird looks up to his left so I do the same, here come the jakes. Four short beards come directly to the decoys, forcing the mature tom to move around front also. Now is my chance, but shaking knees and a befuddled mind make me miss a 15-yard shot at a strutting mature tom, drawing only a couple of breast feathers as it brushed by him. Gosh! He folds, calmly looks around like, "OK, that's enough" and he slowly walks off. I wait for the jakes to leave and gather my arrow (no blood or any gut matter, so I put it in my quiver and spend the next 30 minutes looking to be sure). I gather decoys and start to the truck. That's enough, and I'm still shaking.

I have decided that noise doesn't dramatically spook a turkey, so next time, I will go ahead and slowly open the window giving me a possible shot. I will never again try to stick the arrow through a hole. At almost 70, I am still learning, I hope.

The next morning, I cross the Beaver River and start east listening for birds. Geeze, I have my choice of at least three toms. One is back across the river about ¼ mile away so I ignore him. The other two are spread directly east in what I later find are mature cotton woods. I close to within about 150 yards and set a hen decoy in front. Gathering my gear, I hide in a cedar, load my SHOTGUN with 3-inch 4s which my American side-by-side shotgun patterns through a full choke very well out to 30 yards. I am ready.

I give some soft seductive (I hope) tree talk which truly excites the birds. All three gobble excitedly. I yelp very softly and shut up. I can picture them straining trying to figure where that hen is. A few minutes later, I see dark forms and as the hens fly down followed by the two close gobblers. The third is still talking to me but he moves off in a different direction. I concentrate on the two but can hear the hens giving their normal soft clucks, purrs, and yelps so I do the same. Later, I up the intensity which flat out turns them on. One breaks off from the hens or perhaps they moved on to feed and nest (it's late in the season). He's a smart old tyke as he comes to within 30 yards but stays in a thicket. He gobbles and walks away to 100 yards where he gobbles again, but I still can't see him. He does this three or four times, so the last time he does it, I cut off his gobble with an excited cackle. That does it; he double gobbles a couple times and shuts up. Fifteen minutes later, I assume he is going back to the hens and start to stand up to try to get in front of them, when a red head sticks up 20 yards out. He fans, folds and looks, fans again and is obviously perplexed. He looks at my decoy and tilts his head. (I believe our mature birds are now decoy shy.) He slowly pulls his head down and starts walking off but he comes into the open, I cluck which makes him stick up his head again, and a load of fours ends the ordeal. This is an hour after fly down.

He has a 9½ beard, 1-3/8 inch spurs and weighs 22 lbs. His fan is very attractive and in good shape. Two down and one more to go, but first, my son and grandson are coming in tonight. We have to get them a bird. They show up that night, we have a drink, a great dinner at the lodge, and go to bed.

Next morning, our son Drew, his son Ethan (7 years old), and I are back near the same roost. There are two gobblers in front of us and one apparently alone ¼ mile west. He is a much more silent bird but he does answer my tree talk so he is interested. The birds in front are truly interested.

Following my normal style, I tree talk until full daylight, see them fly down, then start to call seriously. The two in front still have hens, so I am prepared for a long calling session like yesterday, but the gobbler to the left gobbles once more as he flies down. Maybe he will be callable, but I can't tell whether there are hens or not. We call another 30 minutes or so, my son stands up to get rid of a cup of coffee and says, "Oh no" (only more graphically). Slowly sitting back down, he tells me the bird on the left is walking toward the other birds about 100 yards out. I start giving soft yelps, purrs, and clucks.

He does not gobble, but in about 10 minutes a red head appears at 40 yards in a thicket. He tilts his head and looks around. Spotting the decoys he fans, folds, and starts toward the decoys passing by them at 30 yards.

Drew has his shotgun on the bird's head, Ethan is absolutely still, I have tears running down my cheeks. This is my job, and it's the first time I've had three generations all looking at the same bird. The bird clears the decoys, and a load of 4s out of a Winchester SX3 ends his amorous intentions.

The bird has 9½ inch beard, 1¼ inch spurs, a nice fan, and weighs 20 pounds. After taking lots of pictures we walk back to the truck and go to the lodge.

That afternoon we go to a local pond where Ethan demonstrates his fish-catching skills. Starting with a top water lure, he shows he can cast and land fish, so I put on a large plastic worm. He learns to cast, lift, reel in loose line, lift again, etc. When he feels a tap, he gives a rod length of line and sets the hook. He lands a 2-pounder by himself.

Folks, it simply does not get any better than this for a grandpa.

Spurs, Crabs, and Stripers

Kathy and I are originally from the Chesapeake Bay area in Virginia, and we are blessed to still have a home on the water. We have kept a recreational crab pot license over the years, so we try to go back frequently to partake of the bay's bounty. The bay is suffering from people and pollution overload but she still offers much.

First, we need to get a long beard for Ed, my best buddy back east who lives in Deltaville, Va. Just down the road from our place, he has a large farm with lots of birds. The first day he, and I set up about 100 yards from a treed gobbler and start to tree talk. Over the next hour or so, he flies down, gobbles frequently, comes within 30 yards, but holds in a thicket that we cannot see through. Tiring of talking to a stubborn hen, he walks off.

The next morning we are back in a blind with a hen and Jake decoy in front. He does the same thing. He tries that the next morning also, but I used that tactic against him as you will see.

It's raining, so we decide to just sit in the blind and see what happens. We are doing that and solving the world's problems when I pull out the box call and try a loud series of yelps. A gobbler answers way off but he is obviously excited (horny). I use a slate call and mouth call to softly flock talk. He likes that and gobbles excitedly much closer. Continuing that, we see a red head running toward us on a trail. He closes rapidly, fans and starts to strut when he decides that the decoys are not moving. He turns and slowly walks off (once again, I think they are becoming decoy shy) only to be stopped for good with a load of 4s out of Ed's semi.

This bird is a true trophy with 11½ inch beard, 1¾ inch spurs, and he weighs 22 pounds. Ed and I take lots of pictures and walk back to the truck. I shall return alone tomorrow to finish the job.

Next morning, I walk to the same area but go toward the stubborn bird about 50 yards very quietly setting up but not quietly enough. At daybreak, I call softly. He answers only 20 yards away, so when he flies down, he flies off into the distance. I should have known better, but thinking he was spooked, I stood and walked into the woods another 200 yards. Calling, I get two answers. One is in front about ¼ mile, and the first bird is now back exactly where I was (another lesson learned). Deciding to go after him, I move in about 20 yards and start calling. He is responding, but this bird is wary. For the third day in a row, he hangs up at 100 yards. I decide to close.

Calling one more time to establish my location, I get up quietly, or as quietly as a 69-year-old guy can get up, and slowly stalk toward him. Neither of us is talking, so I have to be very careful. There, he sticks his head up and tries to fly but a load of 4s out of the American knocks him down. I don't like to body shoot but, hey, this is the third time for this bird. He dies almost immediately, but when cleaning, I had some shot to remove.

Mine was the dominant bird I think, but he was considerably smaller than Ed's. He had 10½ inch beard, 1¼ in spurs and weighed 20 pounds. Interestingly, his head was much smaller than Ed's and he weighed less. I think Ed's bird was the previous dominant but, like the caller, he was past prime and had been beaten by my smaller bird. I have seen this before and apparently the beard does not get smaller as they pass prime. Mine was a wary bird that challenged me, which I like.

Going back to the house, Kathy and I set out the crab pots and start fishing. We catch a few fish, lots of crabs, and paddle around our beloved Chesapeake for another two weeks.

I have taken three mature birds. The first was a Merriams with his brilliant white feather tips, the second was a Rio with his tan feather tips, and the third was an Eastern with his dark chocolate feather tips. I am one lucky fellow.

Rivers
Metaphors for Life

Rivers flow not by us but through our souls, tying our past with the present, going into the future giving us vivid images of spectacular previous encounters through the present, leading into the future where the vision is not so acute but still enlightening. I see Dad and me over 50 years ago paddling tidal streams in eastern Virginia. He's paddling and I'm sitting in the bow waiting the flush of black ducks, mallards, or a host of other species. Later, we switch and I paddle while he shoots. His shooting was much more accurate and for some reason, my vision is sharper when he is shooting. I'd give a life's accumulation of material things to do that one more time with him.

I see our sons Travis and Drew and me sharing the same experiences on streams in Oklahoma. Now, we mostly are walking streams and sneaking. The species are mostly gadwall and mallards but sometimes widgeon and teal flush. Thank God, we still do that or something similar today.

I see the Arkansas in Colorado where we are set up on a river bank with a stool of decoys out front. The Robo Duck is spinning his wings and we are calling. This is today or at least two days ago where Drew and I are trying to learn to duck hunt Colorado. That day we saw two bald eagles, a whitetail deer and four mulies, one of which is spectacular. We also saw four ducks and two flocks of geese wearing oxygen tanks since they were so high. No ducks fell but that's unimportant.

The future is less clear, but Drew has a son of six who will be introduced to duck hunting perhaps this year. Maybe we'll go to the South Platte where it empties into Antero Lake and try to bring a mallard or two down. Certainly, we'll talk a lot as grandpas and grandsons should do. If he develops a love for things outdoors and nature, that's all that matters.

"Water and meditation are forever wedded." This little quote from Moby Dick attributed to Ahab says it all. I cannot sit by water, let it flow through me flushing the meaningless world away without having all these thoughts. To me, clear undammed streams, especially those that hold trout, have the feeling of cold evenings with warm fires and close friends sharing drinks and stories. If I let it happen, rivers always give me this warmth even when choked with ice floes creaking as they pass by. I always leave this experience completely stressless with a Zen type glow.

The most thought provoking rivers flow around and over rocks and sweepers (limbs hanging in the water) hypnotically gurgling in patterns that suddenly become crystal clear—they talk to me. If possible, I always camp close enough to hear that music, and it never fails: I wake in the middle of the night to conversations that sound hauntingly familiar. I only wish I knew the message, I'm still learning.

Placid pools require polarized glasses to spot the large trout (bass, catfish, red eye, etc.) waiting in ambush with an almost imperceptible "slurp" as life feeds life. Why can't we understand the axiom of nature that one life has to die in order for the other to live? I sometimes lie on boulders and peer straight down into a clear stream to watch brookies feed. Sooner or later they accept me and will lie directly under my boulder waiting to rise for that bug. I have had them stare directly into my eyes communicating the circle of life that "we're not all that different, you and I, and I'm just as important as you." Sometimes, I take my two weight, match the bug and demonstrate that; almost always, I release him/her but sometimes I complete the circle.

A spring starting its flow to the ocean is much like a child starting life. Others streams join the flow, as friends, spouses, and children join our lives enriching us greatly, making the stream much larger. Sometimes a waterfall appears as excitement occurs and sometimes a placid pool shows how contented life can be. Hopefully, the stream and our lives will flow uncontaminated to the ocean.

Rivers and streams are metaphors for life. May they flow clean, unpolluted and healthy forever. They cannot protect themselves; we are the custodians holding the streams for our children and theirs to use. May the spirit that flows through streams bless your life.

An Evening Stroll in Shenandoah National Park

He is humped in the back, trying to look as large as he can to intimidate the other bucks, but his symmetric 10 points are probably enough to accomplish that. G2 is about 14 inches tall, and his spread is around 20 inches. G4 is about 4 inches, so he is a beauty, probably scoring around 170. He knows I'm there with my camera but I'm the least of his concerns right now. He is glassy eyed, he licks his nostrils frequently so he can smell a receptive doe better, rubs bushes periodically trying to get a doe's attention and maybe scare off lesser bucks. It is RUT in Shenandoah National Park (SNP).

I have rendezvoused with this buck three nights in a row so he knows I'm there, he knows I won't bother him, and besides I'm not a good-looking doe. Just across from Big Meadow in SNP is a blocked road that goes to Hoover's retreat, and that's the path I have been taking. After half a mile or so, I have it to myself except for all the wildlife. What a place this is for anyone who enjoys animals.

I slowly walk about 30 yards behind him, moving when he does, and stopping when he does. Any closer and he gives me that look that says "that's close enough." I honor that request or command. He weighs more than I do and he has formidable weapons.

I am trying to take good pictures of this magnificent animal, but he is a master at avoiding the camera. Every time he stops, he is behind a tree or deep inside a shadow. I try a flash to get some fill, but that irritates him and I don't try again. Even with my recurve, it is an easy shot, but he is safe for several reasons including that hunting is not allowed there. Each evening, this partnership between him and me goes on for about 20 minutes before he catches a doe estrogen smell or just plain tires of my company. Without a look back, he trots away and is off looking for more does.

My heart is full like a stream receiving snow melt in May, my step is lighter, and I am grateful for that experience. This would normally be enough, but there may be another creature waiting for me in an acorn patch farther down the path, so I move deeper into the woods.

There, up ahead, is a large black object in the acorn patch. The black bear is back also for the third night in a row. I walk up the path until there is a quiet way to move closer to him. He has been letting me get within about 30 yards, maybe a little closer tonight. So, I walk softly toward him, watching carefully his body language for warning signals, and I do get them later.

He weighs about 250 pounds, and his coat has that deep luxurious appearance of a bear getting ready for winter. There is a small white patch around his throat, but that's it, so he is flawless in appearance. What a trophy he would be but, once again, I'm hunting only with a camera. He is not nearly as shy as the buck and will let me take some pictures even with a flash, but it is so dark that all the camera shows is a black blob. My mind camera has a super sharp image that is impossible to delete.

This is our last night in SNP, so I decide to push it a bit. (Please read the safety footnote, p 168). Watching his body language closely I approach to within about 25 yards by never going straight at him but angling back and forth. There, his body stiffens; he turns and looks at me, woofs one time and pops his teeth a couple of times. That's it. I start to back out, and he slowly ambles away. I'm a bit sorry I pushed it, for I shall miss him.

The walk back out of about two miles is only vague in my memory. Nearly dark, the trees seem to reach out and hug me as I walked slowly. The darkness is like a comforting blanket pulled over my head. I am most comfortable alone in the woods, and this is a very special place. Seeing many deer, I did not even attempt any more pictures or stalks. My senses are already over loaded and any more would be wasted.

Brook Trout and Rutting Bucks

The day before this story, we were hiking along a remote trail beside a trout stream. The crystal-clear water bubbles and tumbles over boulders as it seeks its way to the Atlantic. Simply stated, there are some of the prettiest trout streams on the East Coast in SNP. This one held a great surprise and education for me.

As I approached the pool, I could see brookies having a territorial dispute much like bucks in rut. The larger more aggressive one would charge the smaller less aggressive ones, running them off. Then, the aggressor would return to a somewhat sulking female and attempt to court her. Remembering the frustration of that attempt in my teenage years, I decided to watch more closely.

A large boulder lay on the north side of the stream. The pool was shaded so there was little glare and I could see extremely well. I crawled out on the boulder and lay over the pool with my head only 12 inches or so above the water's surface. Of course, the beautifully colored brookie in all his spawning regalia bolted under the rock. I lay still.

In a few minutes, he stuck his head out and looked me eyeball to eyeball before bolting back under the rock. I again lay still. Next time, he ventured out a bit farther before bolting until finally, he stayed out and went back to his courting displays. I had a front row seat to a spectacular show.

He continued to chase smaller males away while watching me all the time. Then, he would swim up to the female who by now had rejoined the scene. Swimming over some gravel, they fanned the bottom, creating a small depression; then she would release eggs over the hole. He would immediately swim over and release sperm. WOW.

If you've not seen or caught a large male brookie in full spawning colors, you need to. Period. Try just about any fly size 14 or 16. I've had really good luck with Parachute Adams. If the little ones won't leave it alone, try a size 12 or even 10 Adams. If you really want to catch some, put a small dropper nymph (size 18 or 20) below that. What nymph doesn't seem to matter much, but I like small bead heads. TIGHT LINES.

You should try SNP in the fall. If you do, say hi to the buck and the bear for me.

Animal Safety. I have been around animals all my life of more than 70 years, and am very comfortable with their communication. I believe they recognize this as described in my story above. I never approach directly. I never make sudden moves and frankly, with every step I am looking for a tree to duck behind (especially for bucks in rut). The deer let me know I wasn't worth the trouble, and the bear respected me until I got too close. Even then, he wasn't trying to hurt me, just communicate. Woofing and popping are his two communications that you should never ignore. NEVER run from either. I firmly believe that running can trigger a chase or worse. Finally, I never make eye contact with a wild animal. If they look at me, I look away and even turn so my shoulder is pointing at them rather than an aggressive face-on challenge. I love wildlife.

What is G1, G2, etc., and What is a "Score?" Starting with a deer's forehead and moving out, the antlers are named G1 for the first ones on the right and left (brow tines), G2 for the second pair, etc. to the end. Scoring is done by measuring the length of each of the antlers, the length of the main tines, the spread or width of the antlers, and the diameter of the main beam halfway between G1, G2, etc. All these measurements are taken in inches and are added to get the final score. As said in the story, I estimated his score to be around 170 which would place him in just about any record book but not at the top.

A Great Day Elk Hunting
9/16 Monday (I Think)

Yesterday at lunch, not one but two red-tail hawks circled me for about 15 minutes. For those of you who don't know, that means really great hunting luck will follow. This has happened to me twice, and both times I had great days. The Cherokee call him Tlanuwa and they think he is good luck for a hunt. Thus, my hopes were high.

Wayne Turner showed up in camp that afternoon. On most hunts, the first day or two are spent looking for Wayne. Then, he shows up and at that point—on all hunts—I do not hunt with anyone else, other than my sons, Travis or Drew. I want to be alone. Ortega-y-Gasset said something like "What I'm hunting most, when I hunt, is myself."

Therefore, at the last minute, I decide to hunt up a creek alone and just spend the day wandering. We have found elk but not large concentrations; perhaps they are up this creek bottom. The rest go to sit a large water hole not too far from camp. A cup of coffee sitting by the campfire waiting for enough daylight to not have to use a flashlight is about as enjoyable as life gets. At last, there is enough light to see, and Wayne and I start up the creek.

As the dawn appears, there is frost on the mountain and a fog over the stream. The fog gives the whole scene a surreal feel as it moves up and down with the wind. I step into the dark timber which feels warm and comfortable yet a bit forbidding. At this point, the green on my camouflage becomes the moss on a rock, the brown streaks become limbs of a tree and I am only the glue that holds it together. I sincerely feel like I not only belong there but I am part of the woods. That is important and welcome, and the only time I feel that is when I have found Wayne (myself).

Almost immediately, there is elk sign in the form of fresh tracks and scat, but the sign reads like "traveling" elk, not elk that are staying there. I need concentrated sign that says they are hanging out in the area. After still hunting (very slowly walking, reading sign, listening, etc.) for about 1.5 hours, I start to see more concentrated sign as apparently they are funneling together to go to the bedding area.

At one spot, I must have intersected a bull for the tracks are large, extremely fresh and trotting. He stops periodically to look back (maybe for whomever is following), but I never see him. Even though elk can walk faster than I can run, I attempt to speed up in the hopes of at least seeing him. He knows where I am all the time, but I never do see him. They are really great animals.

Tired now and about 2 miles from camp, I eat a candy bar and watch the confluence of four different trails for about 45 minutes. The cold slowly breaks through my outer clothing, destroying the built-up body heat, and I am cold. I start moving again. More and more sign appear as I must be approaching the bedding area. Each of the small feeder streams has dense sign, so I hunt them carefully; but the elk aren't there. At one, there are about 10 totally demolished trees from elk rubbing on them, but the elk are gone.

The start of the stream appears as it forms from several meandering small rivulets that appear out of a bog. This is an excellent site for an elk wallow, but there is not one. The elk must be going on up to tree line to bed. I decide to go on up there.

Crossing the stream, I find an open meadow and climb it to above tree line. The grade is probably 20% so I am literally crawling my way up and have not yet looked at the valley below. There's a rock outcropping to which I make my way. Getting there, I sit down on the rock and turn around.

I've died and gone to heaven, for there below me is one the grandest views I have ever seen. The elevation is about 11,000 feet, and I can see at least 50 miles down toward and past Chromo, Colorado. Because I am on a rock, I can rotate and have an unobstructed view for about 330 degrees (almost a full circle). The camera comes out and I play tourist. Almost a whole role disappears before the clicking stops. This is absolutely incredible, and elk are (temporarily) the last thing on my mind. The view and experience are truly unique, and I may never experience this same view again. (Hopefully, there will be many others.)

I break out lunch and eat. Then, all of a sudden (this will be hard to describe) a very strong feeling of contentment and emotion fills my whole body as I realize this is unbelievable. Wave after wave of emotion floods my whole being, even yanking out a tear or two for no reason other than this is where I want to be at this moment. I am part of this scene, not a stranger infringing on it.

This culminates in my realizing I AM KING OF THE MOUNTAIN. I grab my bugle and let out with the biggest, baddest mean-ass elk bugle I can manage. It may not impress the elk, but it did impress me.

Actually, it did impress an elk as immediately below me a bull responds. He is an immature bull because he does the "aeeeeiiiiii" part very well. When he gets to the "iiiiicccyuch" he chokes. Obviously embarrassed, he tries again and does the same thing. I am back to earth.

I grab bow, arrows, calls, pick up lunch trash, and I'm back to hunting. I chase back down the mountain but never find him. My mature bull call probably scared him so much that he is afraid to answer. Anyway, 30 minutes later with no response I settle back in to hunt and start toward camp which is about 1000 feet down and 3 miles away.

Frankly, elk are the last thing on my mind as I'm drifting through the woods, when suddenly sticks begin breaking and I hear what sounds like Sherman going through Atlanta. About 15 yards away, a spike bull jumps up and runs to a close clearing. "Spike" means he has two antlers (one on each side) is probably 18 months old and is illegal. However, he may not be alone, so I grab my cow call and start sounding like a concerned mother elk (I hope). There are no other elk; but he sure likes the sound of Mom. The next thing I know, he runs back to me and stands there broadside about 12 yards away. Strangely we are comfortable with each other. Even I could make that shot, but he is illegal. We stare at each other awhile and he trots off.

The following occurred on a later hunt. The smell is a pungent yet very pleasing mixture of mint, sage, and wet grass—like a freshly mowed hay field. I stop my climb, close my eyes and just breathe deeply through my nostrils. Gosh, the effect is relaxing and healthy, like waking in the morning after a great night's sleep and just laying there letting my senses also wake. So, this must be aroma therapy; if so, I like it.

But, my long bow, wood arrows and I have another mile and 500 feet to climb before I start hunting. Reluctantly, I leave and start climbing again. My day is made already; what happens now is just icing on the cake.

At a saddle, I stop and start a squeal bugle followed by some cow calling. The elk weren't there but three curious deer were. A doe, button buck and another doe (brother and sister) almost step on me as they seek the source. When I finally say "hi" and they bolt, they are only three steps from me. Nice icing.

I finish the hike back to camp about 2 hours later, chronicle these notes, and just sit there for a long time. With my best buddy, I have just hiked about 6 miles, covered around 1000 feet of elevation, had one of my best days ever, and never drew my bow string. Wow.

Where are those hawks anyway? I really would like to see them again.

Memories of a Life Spent Outdoors

My sun is lowering in the west
> Usually the prettiest time of day and life, it is also the best

Seven decades of a wondrous life I've chased fur and fin
> Sometimes losing but always in the end I win

What a life I've had, most of it outside
> What great memories I have, wanting to lay none aside

Years ago, I chose to live life mostly outdoors while being a responsible adult
> I hope I succeeded but, by damn, **I did it my way**

These are Memories of a Life Spent Outdoors

Hey, I won't change, what life remains will be spent outdoors as I think
> My life is flying by as if in a blink

For the rest, I'll be outdoors, that's what God and Dad had in mind
> When very young, I first sat with them in a deer blind

Annie at my feet, boys grown up, Kathy at my side, I will continue to live with no fear of dying
> I'm going fishing but first, Kathy and I have oysters to harvest from the Chesapeake

What Great Memories of a Life Spent Outdoors

I love you, family (Kathy, Travis, Drew, Dad, Mom, Ken, Annie, Okie, Adam, Sooner, Taz)

Part IV

Five African Animals
All Trophies
(To Me, Anyway)

Wayne C. Turner
Dillon, Colorado 2014

The first ale is half gone as I sit here at Denver International Airport anticipating what Africa is going to be like. Dianne and Ed (two friends from Colorado), Donnie and his daughter Selena (friends from Reno, Nevada) are meeting Kathy and me here at a pub in DIA and leaving for an almost 30-hour trip, counting layovers and flight changes, to South Africa. Then we have a 6-hour ride south to (WOW!) AFRICA southeast of Johannesburg to begin an 8-day hunt. To say I am excited would be an understatement. Kathy and I are experienced travelers, having visited many countries, but this will be our first hunting/photography trip to Africa.

My 58-lb. Black Widow Recurve is broken down in my luggage along with carbon arrows topped by 240 gr Steel Force single bevel broad heads with a total weight of almost 700 gr and a significant forward of center (weight forward heavy) of over 60%. They are small spears traveling slowly, but if they hit, the momentum and penetration are awesome. Later my professional hunter (PH) Benton (Ben) tells me the penetration is impressive as he digs my wayward arrow out of a log. Paper clips are holding my twist in the bowstring, so hopefully my brace height of 8.5 inches will remain and work. When I reassemble later, the brace height is exactly 8.5 inches, although a day later I have to twist a few more times to keep it there. Oh, that weight restricts my shots to about 20 yards so I have to be close, and those African animals are BIG.

The flight is long, boring and very tiring. Luckily, Kathy and I can sleep most anytime and anywhere, so we actually get some sleep and feel pretty good when we land in Johannesburg. We go through immigration smoothly since we have no firearms, and pretty quickly we meet Benton, our professional hunter, in the waiting area. Immediately I like Ben, and we talk quite a bit while driving south. Over the next few days, we spend many hours in blinds trying to get a good bow shot, so we get to know each other quite well.

African Hunt Philosophy

I chose to go on safari in Africa only under conditions that met my personal philosophy, selecting a concession (company) that guaranteed the following:

1. *I would never be asked to shoot from a safari vehicle.* The vehicle was necessary due to the many miles we covered each day. We glassed from the vehicle until we found an animal we wished to stalk, then carefully parked, hiked a distance, loaded the rifles and began a serious stalk. One such stalk took nearly a whole day before the final shot. Never were any shots taken from or even adjacent to the vehicle.

2. *I would be guaranteed that every kilogram of meat was consumed.* We ate as much as we could, and the rest was given to a local school system that gratefully accepted it and ate much better because of those gifts. Of the meat we ate, the absolute best was the zebra tenderloin; I would partake of that again, should the opportunity present itself.

3. *We would take the time to watch, enjoy, and photograph all sorts of wildlife including that which we had no intention of harvesting.* This was our first and perhaps last trip to Africa; we wanted to have a total African experience.

4. *I would be given a few minutes alone with the expired animal to pray over it and to thank it for the sacrifice it made.* This is very important to me.

WOW Africa agreed to all the above and, in fact, those are their standards.

Day One

Into Africa

The first day, we drive to a blind about 8 a.m. to see the "pet" eland, Izak, waiting for us. Eland is not on my list, and I have careful instructions to not shoot the pet. He is a good decoy; maybe he can bring something in. We see nothing else the first morning, so we get a ride back to the awesome lodge for lunch. Our chef, Sidney, is very talented and we enjoy every meal of all sorts of exotic wild game. What is our favorite? I'll surprise you later in this story, for you would never guess. Sidney beats a set of drums for each meal and is quite the musician.

The lodge is immaculate and the staff extremely attentive. It was Ruark or Hemmingway who said something like, "You never visit Africa once." Well, the lodge and staff did everything they could to make that come true, and it probably will.

The first afternoon, we go back to the same blind. Izak is still there; he runs off 100 meters as we drive up and watches us unload. Hearing move-

Chef Sidney and his Stove

ment behind the blind, we look, and there are several kudu. While not on my list, I would likely shoot a nice kudu bull as I think they are spectacular.

The kudu and pet eland stand just outside the blind wanting to feed badly, but the wind is wrong. They try to come in, and I can hear them munching on the grass, but every time they get close, I feel a draft on the back of my neck as the wind shifts, blowing our scent toward them. They bark, jump, and scramble off a couple hundred meters only to do it again. They want in, but our scent is too much. Right before dark they can't take it anymore and decide to come on in. We literally can hear kudu breathing just outside the blind, but they won't come around the corner; they were about 10 meters (at that distance 10 meters is about the same as 10 yards) so even I would probably not miss. Izak gives me a 12-meter broadside shot; tempting, but I honor the instructions to not shoot him. Wind shift, barking everywhere, hooves striking rocks, and then, silence; they are all gone. Good, it's time for dinner.

Day Two

Ben Builds Two Natural Blinds with Local Vegetation

Grabbing the recurve, Ben and I sit in the same blind with nothing eventful other than watching African wildlife all morning as they adroitly avoid my shooting lanes. This is quite a show that lasts until, I would guess, 11 a.m. (I never once took my watch or phone; this is great). Then they all slowly amble off. Ben blows in the walkie talky and a few minutes later, the truck drives up to the back of the blind where we load quickly, hoping to avoid spooking wildlife.

After a great lunch and a short nap, we drive back to another site where we have seen impala and oryx. Ben shows his skills by building another great natural vegetation blind with a porthole shooting lane (remember that lane) designed to give me a good shot at about 15 to 20 meters.

Ben Looks through the Blind

Once inside the blind with a face mask to cover my gray hair, I am practically invisible.

We spend the afternoon in the blind and sure enough, the show starts about 3 p.m. First, impala does and babies show up. Gosh, I like watching them, as they put on quite a show. We watch this for about 2 hours with only small bucks at the scene, but the show is great. At 30 before dark, Ben and I agree that there's no string pull today, so I lay the bow down, quiver the arrow and continue watching until some unseen event tells them it is time to leave. Then, Ben blows in the walkie and the truck comes in. Wow!

Day Three

A Zebra and Oryx Sacrifice Themselves for Me

Kathy and I feel great. We have a wonderful night of sleep, an awesome dinner, and a great breakfast. Izak, the lodge owner and namesake for the eland, lends me a rifle. It is a 7.62 with modern optics. I estimate it at somewhere around a 280 and immediately fall in love with this old WWII sniper rifle that had been cleaned and restored. It never fails me, although I let it down a bit later on. The rifle was zeroed in at 200 meters with very little drop from 200 to 250 meters. I never shoot it beyond 250, so I'm not sure what it would do then. Bottom line is I never adjust elevation, although I did consider windage. They didn't know what the pull was, but I felt it was very light, to the point that it fired before I had a chance to jerk—perfect.

We drove out from camp about 7 a.m. and saw lots of game on the way to Ed's blind. He has stuck with his recurve, so he is back in the blind, but he is antsy. I know he will start stalking soon, bow or not.

After dropping Ed and Johanne at their blind, we drive another 2 km and see many animals of different types, except what we wanted, until rounding a curve. There, directly in the early morning warm sunlight, is a large oryx or gemsbok. I estimate at 35 or 36 inches but Ben says "bigger." Turns out he was right. The light on the oryx was as if he were selected for harvest and a spotlight was trained on him. (Turns out "he" was a "she" but at that distance and light, with these untrained eyes, I could not tell).

We continue driving another km or so until we are completely out of sight. We have the wind in our face and the sun over our left shoulder, which was not perfect but close. To clinch the deal we have a row of trees running from us to her. We crouch and stalk carefully until we see her.

We stalk in a low crouch quietly until we are about 125 meters from the animal; since they are big, this should be an easy shot. Up go the sticks, I go down on them, find the animal in the sights just as she turns quartering away. Quickly, I flip the safety off, get down on the animal, and tell Ben, "I am ready." As she turns going away, he says, "Take her!" BANG—thawk. She drops immediately and kicks once or twice. We stay on her a bit just in case, but five minutes later we walk up. She is dead.

She has 38-inch horns and is quite old. In my opinion that is a perfect trophy, for she is big but really going downhill as evidenced by her bony haunches. I am quite pleased. I spend a few moments in prayer honoring what she just did, and we get ready for pictures.

Ben cleans her a bit and poses her. I am impressed with his photographic skills as he sets her up in an alert pose with good sunlight. Remember, it's still early morning, so warm colors bathe the scene. See the pictures for the results.

We all load her into the vehicle and take her back to the lodge to the waiting skinners. Folks, I know how to skin and butcher game, but these guys and gals are phenomenal. As the cape is removed carefully for a shoulder mount, the rest of the skin is removed with no nicks for a rug, and the meat butchering is started. The skull has almost no meat, so when they are finished, they salt it quickly. I am very impressed and wonder if they will come to Colorado next hunting season.

We have lunch and a short nap before starting off again. This time, our goal is a good blesbok. On the way, we see a very nice impala. These

Oryx and Wayne

little antelope are gorgeous with 18-inch plus horns. We will be back later to build a natural blind.

We see a herd of blesbok with at least one mature male. We drive farther, drop out of the truck, and start our stalk. Getting to within probably 150 meters, Ben glasses, puts his binocs down and says, "We can do better," but about then, we see a fairly small herd of zebra with one large stallion. Kathy really wants a stallion rug, so I whisper to Ben, "If he's good, let's take him."

To Ben's credit, he spends maybe a half hour stalking, glassing and judging. He says, "That one doesn't have good contrast, and his stripes don't run down his legs. I suggest we wait." Then, another zebra appears and Ben gasps audibly. "Now that one is bigger, he has great contrast, and his stripes run way down his legs." He looks at me and says, "I suggest you take him." Getting on the sticks quickly, I have to wait until the stallion separates from the herd. When he does, I tell Ben, "I have him." He quickly says, "OK." Bang—thawk.

A great shot breaks the shoulder and shatters the heart. The noble an-

Zebra and Wayne

imal rears on his hind legs, kicks a time or two, and drops. Staying on him for a few minutes, Ben says, "He is dead; let's go." We walk up.

He is a very nice mature Burchell's zebra with a weight Ben estimates at 750 lbs. (I am impressed at how quickly they can go from metrics to U.S. units). We walk around him and admire the color contrast and the coverage of the stripes. I am pleased, and Kathy will be very pleased.

Once again, I pay homage as Ben gets the truck and drives up. We pose the animal, and since it is only the two of us, I have to help. Those animals are big, but we get it done and take some pictures. This time, to load into the truck, the winch has to do most of the work with us guiding hooves into the bed. It works, and we drive back to the ranch. I am hungry.

By the way, the whole time we are doing this, a blesbok is watching us from about 200 meters. We'll get him or another soon, but first let's go fishing.

Day Four

A Day of Fishing and Watching an Incredible Number of Wildlife

We want to see as much country as possible and experience Africa, so Izak and Linky suggest we go fishing at a local reserve. We say, "Great," so, today is to be spent fishing and watching hippos, crocodiles, wart hogs, pigs and even some Cape buffalo toward dark, wow!

Sardines wrapped around a hook with chicken liver inside and tied together with dental floss make a "sandwich" that the tiger fish and catfish both like. The wire leader joining the hook to line is necessary as tiger fish have the biggest teeth I've ever seen on a fish. We are on Josini Lake in a rented 32 ft. (my estimate) houseboat with nine fisher persons, all from the lodge, including Benton.

The ride from the landing is awe inspiring as immediately we are seeing lots of wildlife. We have a friend croc that is about 3 meters long follow us, then sit off in the distance and watch. We travel past a resident herd of hippopotamuses in the middle of the river, grunting, splashing and diving. The shoreline is teeming with mammals including wart hogs, water bucks, impalas and various other antelopes. This is like a zoo, except we are going fishing.

The catfish bite is light, and they are adroit at picking your bait, but we catch quite a few up to about 3 kg. The bite from a tiger fish needs no announcement; you better hang onto your rod, as they like to grab and run. The fight is also good, and the fish are beautifully impressive with all that weaponry. The teeth are truly impressive.

I catch a sunset picture with reflection on the water to cap the day as we motor back to the landing, tired, but extremely happy. Today I have seen more wildlife than perhaps ever before in my 72 years, but it's not over yet.

Getting to shore, we load in the cars and start up the hill to the highway. There on the right, back in the shadows, is a herd of large animals—Cape buffalo! This is another first for most of us, so we stop, glass with our binoculars and wonder at Africa. What an experience!

We then drive back to the lodge and, folks, Africa at night is dark! The roads are well marked so I am comfortable, but the hillsides are inky black. I wonder what wildlife is watching us as we grind up the plateau to the lodge. Enjoying a nightcap at the lodge around the fire, we share stories, eat another wonderful Chef Sidney prepared meal and go to bed. Tomorrow I go back to the bow.

Hippos in the Lake

Tiger Fish

Day Five

An Impala Jumps the String

Impala are striking little antelope of a tan or tawny color. Their horns are impressive with a desired length for harvest of about 18-plus inches. They don't stand very tall. I estimate that the animals I watch all morning from the blind at about 15 meters distance stand somewhere around 30 inches to the tops of their backs. It is fun watching these nervous little animals dance around almost ballerina like. However, they are all small bucks and does, so Ben and I chuckle as we watch the show. I can't believe I am in Africa.

Show time! A very nice buck of 19 or 20 inches comes in and chases all the young guys around. Finally after an hour or so of this, he starts feeding close to us. I stand to where I can shoot through my little porthole and wait for him to turn broadside. He does and I start the draw (I should have waited until he had his head completely down or turned away); I go through my mental checklist of "straight arm, concentrate, draw, anchor, concentrate, release and follow through." (You see, Fred Asbell, I did listen.)

I watch the arrow pass over his back and lodge in a log right behind him. Thank God it is a clean miss, but I'd rather a quick kill. Ben is laughing. I am breathing heavy and shaking like a leaf. Ben is very good at understanding what that means to us, and he never once shows disappointment. Folks, we had blind sat for nearly 2 days for that shot. What happens? All animals leave, so we sit a while letting the experience sink in and then go to the arrow.

I am using the American standard measurement system for the rest of this story. The arrow is in the log exactly 12 inches from the ground. The log is 4 yards behind the animal who was 15 yards from me when I shot from my 36-inch-high porthole. Using similar triangles, I figure the arrow was less than 18 inches high when it passed over his back. That nimble creature who stood about 30 inches tall dropped so much that his back was under 18 inches. Live and learn. If I ever do this again, I will wait for him to drop or turn his head and go for a heart shot (low) rather than a double lung.

Wayne, Ben and Blesbok

Hey, an old country boy (me) just shot at an impala with a recurve in Africa—something I never thought I'd do. It was a clean miss; he and I both are a little smarter. This is the second best thing that could happen. Thank you, God.

At their appointed hour of 11 a.m. or so, all the animals disappear, and we call the truck for a ride back to the lodge for lunch, a few stories and a nap. Then, we go back to the blind.

We see many types of wildlife but no shots at anything on my list, so we just watch and enjoy. We sit and watch until dark. Back to the lodge for dinner. Yes, and a few drinks needed to tell the story to anyone who would listen.

Day Six

Another Day with the Bow

We sleep in today, so it's nine a.m. when we drive out from the lodge with our bellies full of a Sidney breakfast; that guy can sure cook. We get in the blind and immediately are overrun with wildlife. Before the morning is over, I see two nyala, three eland, and about twelve impala, but the buck never comes back. I pick up the rifle and decide to get serious again with it.

Kathy, Ben, Donnie and I head out in midafternoon and immediately see lots of animals including a blesbok herd. We drive past, get out and start the stalk. Fortunately for them, but unfortunately for me, Kathy and Donnie can watch the whole stalk through the binoculars. We see one nice blesbok buck at 125 meters which is an easy shot—or should be. For some reason, I am out of breath, so the cross hairs don't settle, and I shoot over his back scaring him a bit but not doing any damage. Again, that is the second best thing that can happen since I didn't hurt him. So we hike back to the truck and start riding again.

About sunset, we see in the distance three impala, two of which are bucks. We park, drop quietly out the back, and start a 1-km stalk. Getting to the end of a row of trees, we glass and see he is 18 inches plus, so up go the sticks, I go down on them with the rifle, and he lifts his head to look around, giving me a great broadside shot at about 150 meters. Bang, he drops immediately. This time, I went for the shoulder/heart since it was so late in the day and it works. He never even kicks. The bullet goes in one shoulder, shatters the heart and comes out the other side.

We watch for a while, then walk up to this gorgeous, graceful little animal with large 19-inch horns. I am very pleased. We pose him for pictures, load him in the truck, and drive back to the lodge, arriving there right at dark. A good day.

The lodge prides itself on not wasting meat, so we ate almost exclusively wild game. During our week there, we ate hippo steaks, zebra tenderloin, kudu steaks, stews and sausages made of mixed animals, and wildebeest stew. All of it was tremendous and cooked to perfection. Sidney is a self-taught chef from Zimbabwe, and he knows how to do it. We did have chicken and eggs in the morning along with bacon, but I'm not sure from what animal. Our favorite was, hands down, the zebra tenderloin; what a surprise. All meat that we did not eat was donated to a local school, and it was very much appreciated. No meat was wasted.

Day Seven

Blesbok and Wildebeest

The sun rises early and strong, so I take my coffee outside and sit by the now cool campfire and watch numerous wildlife feed within 1 km. Wow—Africa. The sun is warming on my face and the coffee is good. I love this place and will be back.

We start looking for mountain reedbuck for my rifle and me to pursue but see no big bucks. We do stalk several smaller animals but nothing worth taking. However, from way up high where we are we do see a nice herd of blesbok. We stalk straight at them and they run all the way down the mountain to the valley only to run all the way back up 2 km or so across the valley where they slow down to taunt us. One of the fence builders working up there sees them and walks straight at them so they run back down into the valley where now we can stalk them after we drive back down the bumpy road. I keep the scope cradled in my lap trying to protect it from the bumps.

We drive up into a huge pasture they call the "Serengeti," get out and look for the herd. Immediately we are in grass almost over our heads, so the stalk is fairly easy since we know where the herd is. We stalk to within about 150 meters, and I can see the back and horns of a nice animal. Up go the sticks; I set them high for sight line and wait for him to clear the grass, which he does shortly. Bang. He drops immediately and kicks a couple times. He is a 14-inch blesbok and he was well earned, since we chased him up, down, up, and back down before the stalk. Taking pictures, I am very pleased with my animal. Back at the lodge for lunch, we decide to once again chase a mountain reedbuck; I am very intrigued with these beautiful, dainty little animals. After lunch, we drive back up the mountain to start glassing, and we immediately spot a group. We stalk a couple only to see they are a bit small, so we hike back to the truck and drive back down the mountain. I am getting to know this mountain very well.

Drinks by the fire by 5 and dinner at 7 leaves time for lots of stories, and there are many flying around. Kathy and I, once again, sleep well.

Day Eight

A Wildebeest Challenge

We start after mountain reedbuck again. Driving up and down two mountains, we do see some small ones but nothing harvestable. We also see just about every other kind of wildlife including giraffe, zebra, oryx, smaller kudu, and baboons. WOW!

Enough of bouncing up and down mountains; we stop and talk about our options. I decide I would take one more animal and would like it to be a red hartebeest or large wildebeest. So, we head back to the Serengeti. We see several red hartebeest but nothing we want to take, so we decide to stalk a large wildebeest herd on the far side. Ben and I get out and hike to glassing distance where we see some nice animals, one of which he estimates at a 28-inch spread. Actually, this is an animal Donnie had seen earlier and told us where—thanks, Donnie. Here we go again.

We stalk that animal across the Serengeti, setting up several times only to have the large bull blend in with the herd—a shot we would not take. They bolt, as is typical for them, and stop another km in the distance; we stalk again, etc., until finally they bolt up the hill out of the valley.

It's a long drive, but we drive back out to the road, and drive up the road expecting to see them up high but no—here they come across the road east so we back up and drive up another meadow road to greet them. Getting out, sure enough—there they are. We glass them several times, setting up sticks a couple times only to have the bull mix in with the herd and mill off further.

Finally, they settle down a bit, and the bull separates to look toward us; we quickly set up sticks, I come down and fire. Bang! Thawk! I hear the bullet strike him. It felt good, but then Ben says, "You hit him high." I aimed for the front shoulder, so the bullet missed bone and hit the neck, apparently missing the spine and jugular. So now, I have wounded an animal. Shit!

They run across the pasture and stop on an adjacent hillside; Ben and I stalk across that valley and up the hillside where we have a good view, but we cannot find him (he has a large red spot on his neck). They bolt again, this time at least 2 km up a hill where they watch below for us. We drop into a draw and start the long stalk. By now, Johanne has heard us shoot, and he is glassing from a high vantage point, so he and Ben are talking quietly. Johanne is another set of eyes for us. Donnie gets in the truck and drives toward the animals.

Meanwhile, Ben and I continue the long hot stalk. Arriving to where we can see them, we cannot find the bull, so we continue to glass until they pick up our scent and trot off further. But in so doing, four animals separate and head back downhill. We glass the big herd; the bull is not there, so we focus on the four, and I hear excitement in Ben's voice as he says, "Yes, there he is." Now we have him separated from the herd. The four run across the next valley and hide in some timber.

By now Donnie has driven the truck to beyond them so he heads up the hill to block. Johanne has driven around and is parked at the bottom. His tracker Plata gets out and starts up the hill directly at them. It is now a team effort, and it seems to be working. The four bulls mill around a bit, trying to figure where to go, and here they start up the hill quartering toward us. Ben anxiously glasses and finds the bull. "It is the second one," he tells me, so I turn up the scope, get on the sticks and find the second bull. He runs maybe 200 meters and turns broadside to look downhill. I'm on him at 250 meters. Bang! Thawk! This time Ben says, "OK, don't shoot again, you got him in the boiler room (lungs)." Still, I work in another round and watch the bull make that frantic dying run for 50 meters and then fall over, demolishing a tree. They are big animals.

The other three are running back downhill when I hear Ben say, "Oh shit, you shot the wrong one." Frantically, I get back on the three when Ben says with obvious glee, "Just kidding. You nailed him." With tremendous relief, I turn to him and say, "You asshole." He and I laugh and hug. I really like him.

We walk up with the whole stalk team, and the bull is tremendous with a 28-inch spread, beautiful bosses and huge body. The 250-meter shot was dead center in both lungs; why couldn't I have done that on the first shot?

We take pictures and by now everyone knows I want a few minutes with the animal so they walk off. I apologize to the bull, thank him for his life, and say a brief prayer. We pose the animal and take pictures.

Our last night in camp was wonderful. All of us are exhausted from

Wayne and the Trophy Wildebeest

the long team-effort stalk, but we feel great. The drinks were even better than normal, and the late dinner was perfect. I guess Sidney knew the feeling so he delayed dinner as we talked and drank.

Whoever said "You never go to Africa once" was right!

Wayne C Turner
Dillon, CO
wayne.turner@okstate.edu